1,000,0
s
are availab
at
Forgotten Books
www.ForgottenBooks.com
Read online
Download PDF
Purchase in print

ISBN 978-1-332-04714-7
PIBN 10275522

This book is a reproduction of an important historical work. Forgotten Books uses state-of-the-art technology to digitally reconstruct the work, preserving the original format whilst repairing imperfections present in the aged copy. In rare cases, an imperfection in the original, such as a blemish or missing page, may be replicated in our edition. We do, however, repair the vast majority of imperfections successfully; any imperfections that remain are intentionally left to preserve the state of such historical works.

Forgotten Books is a registered trademark of FB &c Ltd.
Copyright © 2018 FB &c Ltd.
FB &c Ltd, Dalton House, 60 Windsor Avenue, London, SW19 2RR.
Company number 08720141. Registered in England and Wales.

For support please visit www.forgottenbooks.com

1 MONTH OF
FREE
READING

at
www.ForgottenBooks.com

By purchasing this book you are eligible for one month membership to ForgottenBooks.com, giving you unlimited access to our entire collection of over 1,000,000 titles via our web site and mobile apps.

To claim your free month visit:

www.forgottenbooks.com/free275522

* Offer is valid for 45 days from date of purchase. Terms and conditions apply.

English
Français
Deutsche
Italiano
Español
Português

www.forgottenbooks.com

Mythology Photography **Fiction** Fishing Christianity **Art** Cooking Essays Buddhism Freemasonry Medicine **Biology** Music **Ancient Egypt** Evolution Carpentry Physics Dance Geology **Mathematics** Fitness Shakespeare **Folklore** Yoga Marketing **Confidence** Immortality Biographies Poetry **Psychology** Witchcraft Electronics Chemistry History **Law** Accounting **Philosophy** Anthropology Alchemy Drama Quantum Mechanics Atheism Sexual Health **Ancient History** **Entrepreneurship** Languages Sport Paleontology Needlework Islam **Metaphysics** Investment Archaeology Parenting Statistics Criminology **Motivational**

THE AIR AND VENTILATION OF SUBWAYS

BY

GEORGE A. SOPER, PH.D.

MEMBER AMERICAN SOCIETY OF CIVIL ENGINEERS, AMERICAN CHEMICAL SOCIETY, SOCIETY OF AMERICAN BACTERIOLOGISTS, AMERICAN PUBLIC HEALTH ASSOCIATION

FIRST EDITION

FIRST THOUSAND

NEW YORK

JOHN WILEY & SONS

LONDON: CHAPMAN & HALL, LIMITED

1908

GENERAL

Copyright, 1908,
BY
GEORGE A. SOPER

Stanhope Press

PREFACE

THIS volume is the outcome of studies carried on for two and one-half years for the Board of Rapid Transit Railroad Commissioners for the City of New York and, after that board went out of existence, for the Interborough Rapid Transit Company to whom the first New York subway is leased. The work was begun in the summer of 1905 and concluded in 1907.

The original data covering about 2000 pages have never been published, although reports summarizing many of the facts have appeared in the official transactions of the Rapid Transit Commissioners or been read by the author before the Society of Arts of Boston, the New York Academy of Medicine or elsewhere.

At the conclusion of the investigations it seemed desirable to have some of these reports bound together for private circulation, but it was finally decided to put the work into somewhat more extended form and offer it to the general public.

It has seemed desirable to preface the description of the investigation by a few facts concerning the scientific ground-work upon which the solution of problems of ventilation should be based, and to this end the composition of good and bad air, some mechanical principles of the atmosphere and other matters have been included. The object throughout has been to make available in convenient form an account of the essential features of the

178215

investigation in the hope that the information may be of service to persons not necessarily trained in sanitary science but interested in knowing what good and bad air consists in and how to deal with it in subways and other enclosed spaces.

G. A. S.

CONTENTS

CHAPTER I

SUBWAYS AND THE PUBLIC HEALTH

CHAPTER II

CHARACTERISTICS OF GOOD AIR AND BAD AIR

CHAPTER III

METHODS OF VENTILATING SUBWAYS

CHAPTER IV

THE AIR OF EUROPEAN SUBWAYS

CHAPTER V

THE AIR OF THE NEW YORK SUBWAY

CHAPTER VI

THE AIR OF THE NEW YORK SUBWAY (Continued)

CHAPTER VII

HEALTH OF NEW YORK SUBWAY EMPLOYEES

THE AIR AND VENTILATION OF SUBWAYS

CHAPTER I

SUBWAYS AND THE PUBLIC HEALTH

THE development of subways as a means of facilitating the movement of people from point to point within the limits of a single city forms an interesting chapter in the history of modern transportation. Placed in the last few years upon a successful basis as a result of operating and structural devices not before practicable, nearly all the largest cities now possess one or more subways and scores of smaller places are planning them.

In the construction of subway systems, London, the most populous city, has at all times held first rank. Not only were the first subways built there, but, in face of repeated failure, subway construction was persisted in in London until it became a popular success. To-day there are more types of subways in London, and a greater aggregate mileage of them, than can be found in any other city.

EVOLUTION OF THE MODERN SUBWAY

The first important tunnel built solely for the transportation of passengers from one part of a city to another ran under the Thames in London about two miles below London Bridge. The object, as in several other river tunnels which were built in early times, was merely to afford means of passing from one shore to another and not to connect with any surface transportation system.

This first subway was begun in 1824 and was opened in 1843. The cost was about $2,340,000. It was intended for horse-propelled vehicles as well as foot passengers. It was 1200 feet long, 14 feet wide and 16 feet 4 inches high. There were, in fact, two parallel tunnels of these dimensions, running side by side, separated by a masonry pier 4 feet thick. The undertaking paid so poorly that the receipts were scarcely sufficient to meet the cost of repairs. This road now forms part of the East London Railway having been purchased in 1865 for less than one-half the original cost.

The first subway to be provided with its own vehicles for transporting passengers was constructed in 1863 under the Thames near the Tower Bridge. This tunnel was a circular iron tube composed of segments bolted together and was 7 feet in diameter and 1350 feet long. As in modern tube railways, passengers were taken up and down in elevators through shafts. These were 10 feet in diameter and 60 feet deep. Transit from one end of the tunnel to the other was accomplished on a single track by means of cars hauled by wire ropes. This subway was closed to passengers in 1897 and is now used for a gas main.

In view of the failure of these two tunnels to meet popular favor, it is interesting to note the complete success of the Blackwall tunnel which runs under the Thames about six miles below London Bridge. It was opened in 1897 and is used by passengers and pedestrians. The total length is 6210 feet, of which 1740 feet are approaches. This tunnel is next to the largest shield-driven tube in existence, being 27 feet outside diameter. The inside is lined with light-colored glazed tiles and is paved with granite blocks on the inclines and with asphalt elsewhere. It is well lighted and the air is agreeable, ventilation taking place through large shafts on the two opposite shores.

The first extensive underground railway which can properly be termed a system was the Metropolitan and District of London. Its object was to carry the public more expeditiously and comfortably from place to place within the city than could be accomplished on the surface. This road was opened from Paddington, the terminus of the Great Western Railway, to Farringdon Street in the business district, in 1863. Extensions were numerous up to 1884, at which time the system had been so developed as to form a complete circle around the inner portions of the city.

To-day, the Metropolitan is almost the only example in London of a shallow railroad subway, that is, one built near the surface of the ground and reached by stairways. All the rest, and there are six more, have been built in the London clay at depths which vary from 40 to 150 feet below the surface.

The pioneer deep tube subway under city streets was the City and South London and was opened for traffic in 1890. It is about three miles long and, like practically all deep-lying roads, is composed of two metal-lined tubes running side by side. This road has been very successful, carrying in the first year of operation about 2,400,000 passengers. It was the first important city subway to be operated by electricity. The original intention was to use an endless cable for moving the trains.

New London tubes. By far the best known and one of the most profitable deep subways is the Central London which was opened in 1900. After this came the Great Northern and City, opened in 1904, the Waterloo and City, the Baker Street and Waterloo, the Great Northern, Piccadilly and Brompton and the Charing Cross, Euston and Hampstead.

The average internal diameter of these deep tubes is 11 feet 6 inches. They occasionally follow steep grades and curve about continually. The different roads are frequently connected at stations by passageways for foot passengers.

Paris subways. Following the example of London, the city of Paris, in 1898, laid out an elaborate plan for subways. The road was built to relieve the lack of public transportation facilities in Paris itself and to develop poorly populated and distant quarters in the suburbs. It was at first thought that only a part of the whole plan would be carried out, but the success which followed the opening of the road was so great that it was at once decided to finish the entire system. Passengers were first carried in 1901.

Unlike most of the London roads, the Paris subways belong to the type which runs close beneath the surface of the ground, are built of masonry, have two or more tracks and are reached by stairways from the streets. Since the opening of the Paris system, extensive construction has been in progress both by the city and by private capital, so that, at the present time, Paris has the second greatest aggregate underground railroad mileage of any city.

Berlin subways. The city of Berlin is provided with a combined subway and elevated system which was opened for traffic in 1902. It is one of three roads which run partly overhead and extends through the central part of the city from east to west. The tunnels were constructed under the streets, the top sometimes approaching to within 2 feet of the surface. The subway system was considerably extended in 1907.

Other European subways. The city of Budapest has an electric underground railway about two miles in length opened in 1896. It is a double-tracked, overhead trolley system built near the surface of the ground and in design resembles the subways of Boston, New York and Philadelphia.

Glasgow has a subway six miles long partly of the deep, and partly of the shallow, type.

Many other important subways now exist in Europe. Some belong to the period of earliest construction, while others are of a newer type. Nearly all are operated by electricity.

American subways. Aside from the terminal connections of trunk line railroads, the first underground city railroad in America was built in Boston and was begun in 1895. This road runs under the business part of the city with a branch under the harbor and accommodates several lines of trolley cars as well as trains of the Boston Elevated System. It is not provided with its own cars but is used by outside lines which otherwise would run on the surface and interfere with street traffic. It was estimated that at least 50,000,000 passengers used the original subway during the first twelve months after its completion and its use has apparently increased since then by over 60 per cent.

The New York subway was opened in 1904. It was considered to be the most perfect example of subway construction yet afforded and was expected to play an important part in relieving the congestion which is rapidly growing at the southern end of Manhattan Island. It is, like most subways, chiefly of the shallow type. A description of its principal features will be found beyond. (See Figs. 1 and 2.)

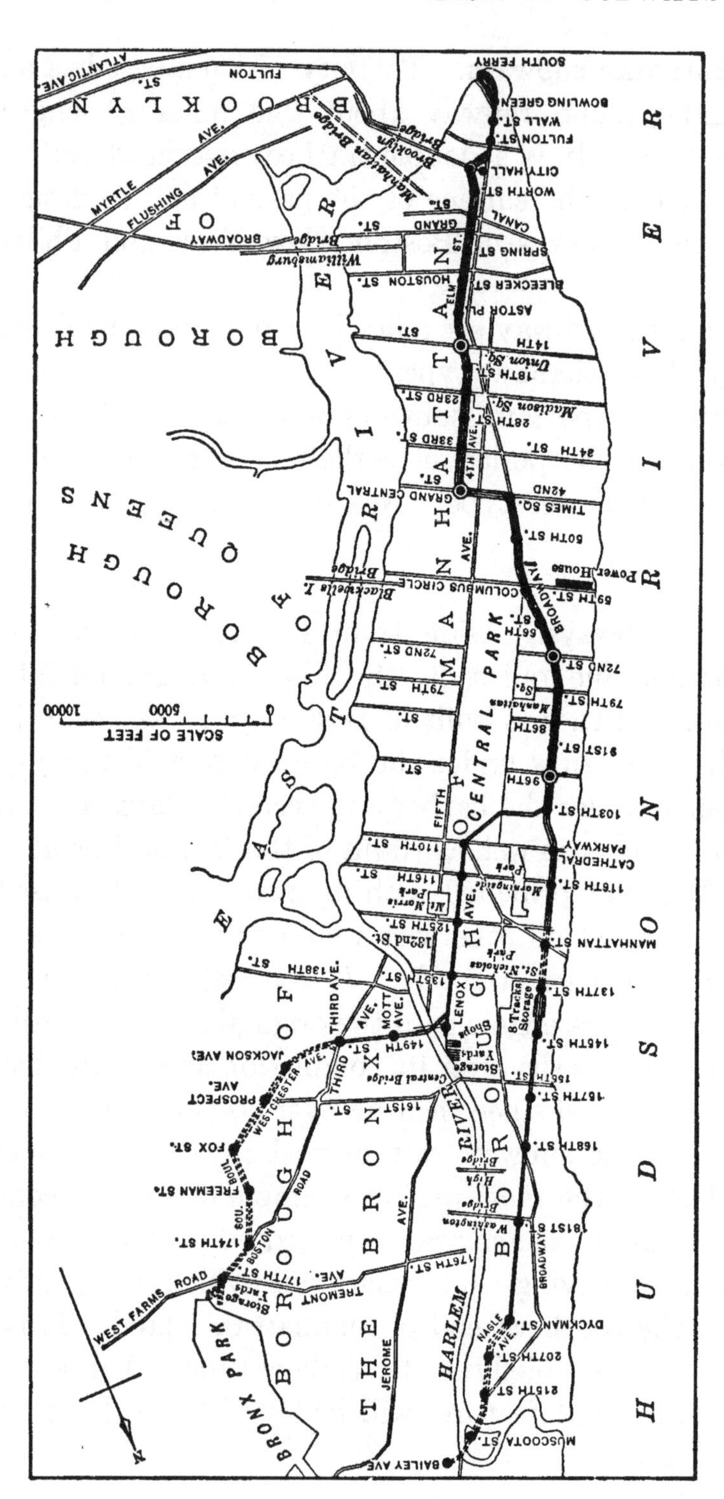

Fig. 1. Route of the New York subway

An account of the plans for the development of underground and under-water transportation in New York which have followed the opening of the first subway would pass the limits of space of this volume. They include roads under the Hudson river and East river and extensions of the original line aggregating many miles in length.

The history of the construction of city subways shows that although underground roads were built over fifty years ago, it was not until electric traction became practicable, bringing with it the possibility of good air, that city subways on a large scale were successful. It is interesting to note that the key to success depended largely upon the question of sanitation. Subways could not be made even tolerably agreeable so long as it was necessary to light them by oil and propel the cars by means of coal-burning locomotives. To such conditions no public could be expected to become indifferent. On this point the history of the Metropolitan of London is conclusive.

Underground adjuncts of subways. In the rapid growth of subways which has here been outlined, the construction has not been confined to the simple form of structure originally adopted. Both with respect to main lines and auxiliary passages, developments have been made which are of considerable interest when viewed from a sanitary standpoint.

The New York subway differs from most foreign roads in having four tracks instead of one or two, in being provided with express as well as local services, in having underground stations with elaborate toilet room facilities and in having numerous underground connections with office buildings and shops. These lateral connections sometimes run for considerable distances through the basements of sky-

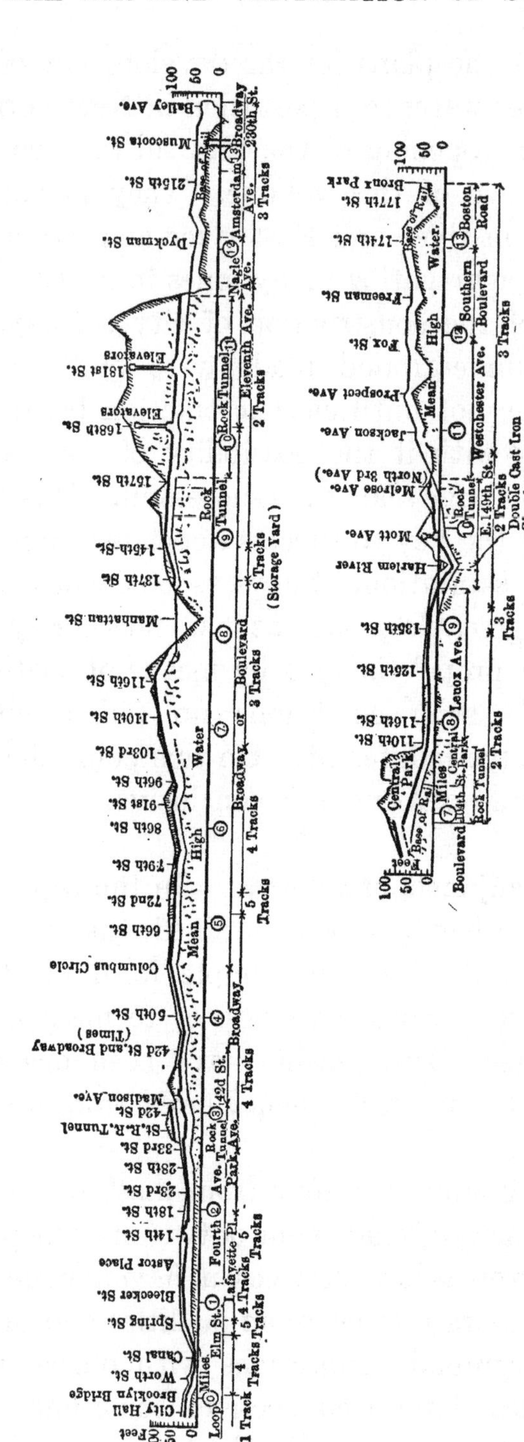

FIG. 2. Profile showing depth of the New York Subway beneath the surface of the ground

scraper office buildings, hotels and shops. The passages often accommodate barber shops, restaurants, news stands, fruit and candy stands, soda fountains and flower booths.

The lateral underground passages of the deep London tubes are often long and devious, but they are used only by pedestrians and do not contain shops. On the station platforms of practically all subways are automatic vending and weighing machines. The articles sold are innumerable in variety.

It is needless to remark that in all these lateral subterranean passages the sun never penetrates and it is always night. Fresh air enters only through doors and through elevator shafts. The air is usually cool. The volumes of air passing are often very large. Where candy, fruit and other food is sold it is usually quite unprotected from the dust and air.

Effects of subways on congestion of population. It would be interesting, if space permitted, to discuss the effect on congestion of population which the construction of underground railroads for urban intercommunication has produced. It would be seen that the ultimate effect has been quite opposite to that sometimes intended, for, instead of relieving congestion, they have increased it. If the subways have given new outlets from the overcrowded sections of cities to the more sparsely settled ones, they have also provided means by which the overcrowded places can be more rapidly reached than formerly, with the result that the worst places have been still further congested. This added congestion means serious inconvenience and possibly injury to the public health.

Some of the inconvenience which results from congestion is but too evident to every person who visits the business districts of a great city during business hours. The streets are so crowded that pedestrians overflow from the sidewalks upon the carriageways and there is so little room on the carriageways that vehicles are able to thread their way only with difficulty and at greatly reduced speed.

Nor does the effect of this crowding bear only on the mere convenience of the public. It interferes seriously with the conduct of trade. This means loss of money. Sir John Wolfe-Barry [1] and Mr. R. G. H. Davison have calculated that the loss of time experienced through the congested condition of London streets affects persons whose annual aggregate earning capacity is £173,291,000 and of vehicles whose time is valued at £16,562,000 annually. If the loss of time to each one of these wage earners is only five minutes in each day of eight hours, the total loss in money which would be produced in a year would be £1,898,534. It is not improbable that the actual loss is much greater.

Careful studies in America and Europe as to the causes and remedies for overcrowded streets have been carried on by a Royal Commission on London Traffic, and an advisory board composed of Sir John Wolfe-Barry, Sir Benjamin Baker and William Barclay Parsons, engineers of international reputation, have shown in a report to this commission that by bringing more and more people from the immediate suburbs, the development of city railways has increased congestion to such a point that it is now necessary to widen streets, plan for proper sub-structures beneath the streets and regulate the heights of buildings if more serious consequences to the commercial welfare of the city are to be avoided.

[1] Memorandum by Sir John Wolfe-Barry, K.C.B., and Mr. R. G. H. Davison, M. Inst. C. E. in Appendix to Report of Advisory Board of Engineers to Royal Commission on London Traffic, pp. 675–6, Vol. 8.

THE BEARING OF SUBWAYS ON HEALTH

A study of all the effects of subway conditions on public health would form an extremely difficult inquiry and obviously can only be dealt with here in a most general manner.

Vital statistics are not available to show the state of health of persons who use subways as compared with those who do not, nor are facts at hand to indicate differences in public health which can, with certainty, be ascribed to underground railways.

For practical purposes, it is probably safe to assume as a groundwork of inquiry that if subways produce evil effects upon the health of the traveling public, these effects are due to overcrowding or some harmful quality of the subway air.

CHAPTER II

CHARACTERISTICS OF GOOD AIR AND BAD AIR

THE ATMOSPHERE OF THE OPEN COUNTRY

So far as known, there is only one constituent of the atmosphere which is directly useful to living beings and this is oxygen. The other gases are not immediately necessary, although they perform many indispensable functions in the economy of nature.

The essential act of respiration is an absorption of oxygen and a production of carbon dioxide. It is natural to think of this exchange as one which takes place only in the lungs, but it is really common to all the myriad cells of the body. The lungs act simply as central exchange depots; the distant cells are the real laboratories.

The principal gases of the atmosphere are present in a mechanical mixture and not in chemical combination with one another. A portion of the oxygen may be abstracted without affecting the other constituents and other gases may be added without affecting the oxygen. Oxygen has always the same properties whether it is freshly prepared or not. Fresh air is simply air which is uncontaminated.

Were it not that the atmosphere is being constantly and vigorously agitated, and an incessant mixture taking place through the operation of the physical principles known as diffusion and convection and the action of winds and lesser air currents, human existence in cities and houses would be impossible and life on any part of the earth's surface would cease to exist.

The reserve supply of pure air. Inasmuch as heavy draughts are made upon the supply of oxygen by living creatures and by the combustion of fuel, to say nothing of the impurities produced, it is evident that the supply in inhabited places must be continuously renewed. The open country, the sea and the upper atmosphere are the great reservoirs of pure air.

How great is this reserve is not certain, for the height of the atmosphere above the earth is not known with exactness. It was formerly supposed that it extended to a depth of about fifty miles, but observations of meteors, which are believed to grow luminous through heat generated by friction with the air of the earth, have led to the belief that the atmosphere extends even to a greater height than one hundred miles.

It has often been suggested that the oxygen of the atmosphere might, in time, become exhausted, and many ingenious calculations have been made with respect to this subject. For example, it has been calculated by Remsen that, assuming the population of the earth to be 1000 million human beings, the quantity of oxygen used in respiration in a year amounts to about $\frac{1}{3,800,000}$ part of the supply. Supposing that the quantity of oxygen required for other purposes is nine times this, then the total amount used up in a year would be only $\frac{1}{380,000}$ of the whole supply. In eighteen hundred years the decrease in the amount would be only 0.1 per cent.

Whether there has been a decrease in the past it is not possible to say. From many considerations it appears probable that the quantity of oxygen in the air will never become appreciably reduced.

Gases of the normal atmosphere. It was supposed until a few years ago that all the essential facts concerning the identity of the gases of the atmosphere were known, but a number of new gases have recently been discovered. The proportion of these new elements is very small and they appear to have no influence upon health.

In the open country, upon the ocean and at all elevations, the proportion of the leading constituents of unpolluted atmosphere are nearly constant. When decided differences have been reported it is probable that they have been due more to errors inseparable from the analytical technique than to real inequalities.

Still, inasmuch as small differences frequently do occur in cities and other occupied places, it is desirable in studying a problem of ventilation to determine the condition of the surrounding outside atmosphere at the time and place in question as well as the atmosphere of the enclosed space under consideration.

To get a general idea of the principal constituents of the free atmosphere under average conditions, we may accept the figures adopted in a recent English Government inquiry concerning the ventilation of workshops.[1]

According to this authority, pure atmospheric air, free from aqueous vapor, may be taken to have the following composition by volume.

	Per Cent
Oxygen	20.94
Nitrogen	78.9
Argon	0.94
Carbonic Acid	0.03
Helium, Krypton, Xenon, Hydrogen, etc.	Traces
	100.00

[1] Report of a Parliamentary Committee on the Ventilation of Factories and Workshops, London, 1902, p. 93.

CALIFORNIA

Included in this list, although not specifically mentioned, are gases of decomposition, principally ammonia and sulphur compounds, which, with varying proportions, are probably always present in minute amount.

According to Ramsay,[1] the rare gases mentioned in the foregoing table and recently separated from nitrogen are present to the following extent; Helium, 1 part in 245,300 volumes of air; neon, 1 part in 80,800; argon, 1 part in 106.8; krypton, 1 part in 20,000,000; xenon, 1 part in 170,000,000.

Many authorities doubt whether that active form of oxygen called ozone actually exists in the atmosphere, notwithstanding the popular belief in its presence and the fact that some investigators have made systematic determinations of what they have thought was ozone for many years. On the other hand, Ramsay is of opinion that peroxide of hydrogen, an active gas resembling ozone in some of its properties, is generally present in very small quantity. The presence of ozone or peroxide of hydrogen is not without sanitary significance in a sample of air, for, owing to their energetic oxidizing properties, the presence of either gas may be taken to indicate the absence of organic matters.

Odors. Contrary to general belief, pure air is odorless, colorless and tasteless. Many of the odors noticed at the seaside and in the forest are due to substances thrown off in the growth and decay of vegetation and are by no means directly beneficial to health.

Odors are nearly always due to gases and not to solid particles, as is popularly supposed. The impression of freshness noticeable in the country and after showers is

[1] Gases of the Atmosphere, London, 1905, p. 257.

more often due to absence of odor than to the presence of it. Even the dirty air of streets may seem sweet and fresh on emerging from a badly ventilated enclosure.

The sense of smell differs greatly among different persons and under different circumstances. Warmth and moisture favor the detection of odors.

To be accurate, what we call smell is a mental effect due to the excitement of the olfactory nerve when suitable substances come in contact with the external cells to which this nerve is attached in the more remote parts of the nasal cavity. Odors can be, and are, continually smelled through the mouth as well as through the nose, many so-called flavors used in the preparation of food being, in reality, substances producing odors and not tastes. Sensations of taste originate in the tongue and are restricted to such impressions as those of sweetness and sourness, bitterness and salinity.

The first moment of contact gives the most acute sensation of smell; this sense rapidly becomes blunt after continued exposure. To most persons a subway, which at first seems to have an extremely strong and unpleasant odor, has no odor at all after five minutes or so. This acquired dullness toward one odor may not affect the keenness of perception for others. It is remarkable how soon persons get used to the odors which in one way or another they themselves produce and object to much less pronounced odors from other sources.

A partial or total abolition of the sense of smell may occur as a result of a catarrhal condition or the action of injurious gases. It is not uncommon for persons who live in cities to become very defective in the acuteness of this sense.

Many gases are detectable by smell when in extremely dilute form. Then the odor of musk can be detected for many years without any sensible loss in weight.

It is doubtless chiefly in their power of suggestion that certain odors of the atmosphere appear to be beneficial. And we may say on the converse side of this proposition that odors which are unpleasant are probably harmful only because of the unpleasant mental impressions which they create. No other sense is capable of calling up such vivid mental pictures as does the sense of smell. Probably no sense is so capable of leading one astray.

Carbon dioxide. Aside from oxygen, the two gases which are of greatest interest in studies of ventilation are carbon dioxide and water vapor. No other normal gases of the atmosphere vary so much as do these two and none, under ordinary circumstances, has such an effect upon comfort.

The peculiar interest which attaches to carbon dioxide lies in the fact that it is a convenient measure of the extent to which air is vitiated by breathing, by lights and by fires. It is a product of respiration and of the combustion of fuel and at the same time varies little in uncontaminated air. There appears to be little difference between the amount present in pure country air, whether in forests or over the ocean.

From accurate analyses made in France by Reiset [1] who used samples of air measuring 525 liters each, 2.96 volumes of carbon dioxide per 10,000 volumes of air appears to be close to the normal for the country. Angus Smith reported that the air among the Scotch hills contained 3.36; J. S.

[1] Annales de Chimie et de Physique, Vol. 26, 1882, p. 198.

and E. S. Haldane,[1] 3.0, and G. F. Armstrong, 3.13 for the country air of Scotland. Three parts may be accepted as sufficiently correct for most purposes.

Water vapor. Water vapor, not to be confounded with moisture in visible form to which the term vapor is so often applied, is always present in a normal atmosphere.

It is usual to refer to the presence of water vapor as humidity and to the ratio of the quantity present to the maximum quantity which could be present at a given temperature as the precentage of relative humidity. Close to the ocean the relative humidity is generally in the neighborhood of 90 per cent, but in very dry places it may go below 10 per cent.

When there is present all the water vapor which is possible, the air is commonly said to be saturated, a phrase which some meteorologists object to on the ground of literal inaccuracy. It would be more scientific to say that the space under consideration is saturated, for if any additional evaporation should take place into it some of the vapor already existing would have to return to the liquid state. But it is customary in this, as in some other directions, to employ the term which is generally understood rather than to insist upon excessive accuracy. Saturation of the air is a sufficiently correct expression for popular purposes.

One of the most important facts to remember concerning the existence of water vapor in the atmosphere is that the greatest amount possible varies decidedly with the temperature. While at 0 degrees Fahrenheit, the greatest possible weight of a cubic foot of vapor is 0.54 grain, or 35 milligrams,

[1] Physiological Magazine, 1890, p. 306.

at 80 degrees Fahrenheit, the maximum is 10.95 grains, or 709.56 milligrams.

The absolute humidity is the actual amount of vapor present expressed either in terms of its weight in a given volume of air or in terms of its expansive force. This expansive force is often called vapor tension. Its amount can easily be calculated when the relative or absolute humidity, temperature and pressure are known.

The dew point is the temperature at which condensation begins. It is simply a convenient way of expressing the amount of vapor present.

Optimum conditions of moisture, cold and heat. The physiological effects of moisture have much to do with health. They control the functions of the skin and so affect other important organs of the body. According to the amount of vapor present in the air, the skin gives off more or less moisture, and, owing to this evaporation, more or less heat. When the air is humid, a hot atmosphere is most oppressive. If the air is cold and damp the heat of the body is abstracted rapidly from the exposed surfaces, and we feel chilled. Dry air is generally the most healthful and comfortable at every temperature. Draughts and winds produce cooling effects by hastening evaporation, by forcing warm air out of our clothes and by abstracting heat by conduction.

At a temperature of 60 degrees Fahrenheit, when suitably clothed and not working too hard, we feel comfortable in an atmosphere saturated with watery vapor, that is, while rain is falling, but from the moment the temperature rises so that refrigeration by the rapid evaporation of sweat from the surface of the body becomes necessary to preserve the normal body temperature of 98 degrees, we become uncomfortable. If the temperature falls much below 65 degrees, we feel cold, unless we exercise. The most comfortable temperature in which to remain without exercise is between 65 and 70 degrees.

All things considered, cold weather is better than hot weather for the average person, for, since more heat is lost from the body in cold weather, the production of heat must be increased and this means more active metabolism and probably greater activity of all the tissues of the body. This stirring up of the vital powers is thought by many persons to account in large measure for the greater vigor of northern, as compared with southern, people. It is even believed by some to increase personal resistance to disease.

Rapid and decided changes in temperature are universally considered dangerous, particularly when the temperature is falling below 60–65 degrees.

Dust. The atmosphere over both land and sea normally contains solid matters in the form of dust, but in uncontaminated air the amount is usually very small and the particles exceedingly minute. These particles are raised by winds from the earth as dust and from the waves of the ocean as fine droplets, the water of which evaporates, leaving the mineral matters in solid form.

Under the microscope, dust from the air in the country has been found to contain particles of wood fibre, fragments of plants, bits of animal hairs and feathers, silica, clay, and, in fact, remains of nearly every living and inorganic thing which is to be found on the earth's surface.

Sea air has been found to contain minute quantities of iodine, possibly in organic combination, and due probably to minute organisms or fragments of once living animals and plants from the sea.

In quiet air solid matters settle toward the earth, but there is, apparently, no limit to the length of time that the finest particles of solid matter may remain suspended in the moving atmosphere; the period certainly covers months and probably years.

The dust from some volcanic eruptions has been known to encircle the earth. Sand, raised by wind storms on the Desert of Sahara, is sometimes carried across the Mediterranean Sea and deposited in the Islands of Malta, in Sicily, and even in the South of France.

While it is probable that the dust normally present in good country air does not appreciably injure health, excessive amounts of some dusts, such as that from the Sahara, affect the mucous membranes and produce symptoms strongly resembling some of those which are characteristic of influenza.

The ultra-microscopic dust contained in the upper air of the open country is an important ingredient of the atmosphere when considered from a climatalogical standpoint, for it exercises an influence upon the temperature of the air, probably determines its color and the illuminating effect produced by the sun upon the sky. With absolutely no dust in the atmosphere there would be no clouds, no twilight and no rain. By furnishing nuclei, the fine dust of the atmosphere permits water vapor to condense and form particles of moisture which result in fog, mist and rain.

The number of these ultra-microscopic dust particles in the free air of country and sea, as well as in the air of cities, has been made the subject of careful study by Aitken and others. Aitken's observations led him to estimate that one cubic inch of air in the open country contained 2000 dust particles; in a city 3,000,000 and in enclosed spaces occupied by human beings 30,000,000.[1] Most of these particles are ultra-microscopic.

Forms of life. Finally, as ingredients of normal air, must be mentioned particles of organized matter, some of which are living, others existing in a resting stage and still others dead. The most interesting of these are the minute forms of life which are almost everywhere present in the atmosphere and collectively termed microörganisms. They include bacteria, yeasts and molds.

By far the most important of these living particles, from a physiological standpoint, are the bacteria. Bacteria of some kinds are almost everywhere present in the atmosphere. The kinds which are popularly called disease germs are exceedingly rare except in the vicinity of living beings.

Mere numbers of bacteria are generally considered by sanitarians to be of little significance. Yet very large numbers are always present in the air of unhealthy places, while small numbers generally occur where the air is pure. It is difficult to place a limit to the number which may be present in air without causing it to be regarded as impure. Some authorities, as Flügge, for example, have suggested 100 per cubic meter as a fair standard.

Unlike dust particles with which they are often associated in air near land, bacteria are found only rarely in the atmosphere over the ocean and on mountain tops.

The sea and, in fact, all wet or moist surfaces, as well as

[1] Nature, Vol. 31, p. 265, Vol. 41, p. 394. See also "Cloudy Condensation," by John Aitken, F.R.S., Proc. Roy. Soc. Lond., Vol. 21, pp. 403–439. Also Smithsonian Reports, 1893.

rain and snow, perform a useful service in removing solid matters from the atmosphere. Contrary to the popular idea, there are fewer bacteria in the air when the ground is wet with slush and snow than when it is dry. Once in contact, dust and bacteria become instantly attached to whatever is wet and do not leave such surfaces unless driven away by spraying or splashing. This is an important fact; it accounts for the observation that ordinary air from the mouth and nose does not contain bacteria, while the invisible droplets given off in sneezing, coughing, spitting and speaking always contain large numbers of them.

Among natural conditions which destroy bacteria are (*a*) Sunlight: Sunlight destroys most bacteria better than many artificial disinfectants; (*b*) Drying: Although the germs of tuberculosis and of some other diseases withstand drying for months, most germs cannot live without moisture; (*c*) Air Currents and Gravity: These are useful in so far as they transport bacteria to places where they are destroyed.

Practically no microörganisms which grow and multiply in nature apart from human life are dangerous to breathe, although certain molds and bacteria grow outside of the human body under conditions which have led some persons to conclude that they are natural enemies to the human race.

Bacteria are divisible into two great classes: (1) The Saprophytes: These live on dead organic matter and (2) The Parasites: These live within, or upon, a living host. Some belong to both classes but show a distinct preference for one form of existence or the other. The various species of these two groups are believed to number thousands and each is as separate and immutable as are the separate species among larger plants.

THE AIR OF TOWNS AND CITIES

It is a well known fact that the air of cities always differs somewhat in composition from the normal atmosphere of the country. The great consumption of fuel and the production of innumerable gases and dusts from industrial establishments give to some cities an atmosphere which is unmistakably unlike that of the neighboring country.

Air of London, Paris and New York. London is probably the most conspicuous example of a city which modifies its own climate. Repeated investigation shows that the air of London contains more carbon dioxide than uncontaminated country air, very much more dust and more microörganisms. There is, moreover, much less sunlight and a more even temperature in London than occur outside of the city limits. The frequency and density of the fogs which characterize the city are largely due to the smoke produced; this fog, acting like a blanket, keeps out sunlight and air.

Angus Smith, who found 3.01 volumes of carbon dioxide per 10,000 in the parks of London, found 3.80 in the streets. His average of 35 analyses of samples from different parts of London was 4.39. The same investigator reported 4.03 for the air of Manchester, 4.14 for the air of Perth and 5.02 for Glasgow.

The most careful, long series of analyses of London air yet made were carried out by Russel who, as a result of 159 determinations, announced 4.03 as the average from the center of London. Butterfield, who was instructed by a committee of the House of Commons to determine as nearly as present scientific methods would allow, the condition of the air in the debating chamber of the House, when in use, found the outside air of London contained 3.37 parts of carbon dioxide, with a maximum of 3.64.

The municipal laboratory of Paris[1] has kept records of the amount of carbon dioxide in the air of Paris since 1890 with results which have been published in the official journal of the Montsouris Laboratory. In the center of Paris the annual average has been slightly higher than in the southern part. In the center it has varied from 3.06 to 3.44. In the south the minimum has been the same as in the center, but the maximum has not been so high. The monthly range has been 3.04 to 3.36 in the center. In both sections the greatest amount has occurred in November and the least in July.

For ten years from 1891 to 1900, analyses were made by the Montsouris Laboratory to determine to what extent the carbon dioxide varied during the day and night. Six samples were analyzed each day between 6 A.M. and 6 P.M. and six between 6 P.M. and 6 A.M. The results showed very slight differences in the southern part of the city, but there was much less uniformity in the central part. The figures follow:

Carbon dioxide in the air of Paris during the day and night, 1891–1900, inclusive.

	Montsouris Park	Center of City
Day	3.10	3.34
Night	3.18	3.21
24 hours, average	3.14	3.27

In the observations made by the author for the Rapid Transit Commission of New York and recorded graphically

[1] Annales de l'Observatoire de Paris, VI., 1905, p. 322.

elsewhere in this volume the amount of carbon dioxide was found to vary from hour to hour in the air of the streets of New York very much as it varied in the air of the subway. It was highest at 6 P.M. and lowest between 1.30 and 3 A.M., with certain fluctuations through the day. The average of 309 analyses was 3.67.

Unwholesome gases. The gases from coal produce an amount of pollution which should not be overlooked. The quantity of coal consumed and the volume of offensive gases given off in the large cities is very great. In New York, in 1905, it was estimated that 9,000,000 tons of anthracite and 6,500,000 tons of bituminous, or a total of 15,500,000 tons of coal were burned. Of this amount Manhattan Island probably used nearly one-half. This coal produced in burning about 3,000 tons of sulphuric acid and 75,000 tons of carbon dioxide per day. The heat from this coal was sufficient to have raised and kept raised the temperature of the air over the whole area of the city. If these gases were not carried away by the winds, but became mixed with the air of the streets and buildings, the atmosphere would have become so poisonous as to kill all the inhabitants.

The atmosphere of cities and towns where manufactories exist is often contaminated with unwholesome gases of industrial origin. Among the gases given off from industries common in large cities are the following given by Notter.[1]

Hydrochloric acid from alkali works; sulphurous acid and sulphuric acid from copper works and bleaching operations; sulphuretted hydrogen from chemical works, especially those which produce ammonia; carbon monoxide

[1] Theory and Practice of Hygiene, 2d Ed., 1900, p. 155.

from iron furnaces — this may amount to from 22 to 25 per cent, and from copper furnaces from 15 to 19 per cent; organic vapors from glue refineries; bone burners; slaughter houses, knackeries; zinc fumes from brass foundries; arsenical fumes from copper smelting; phosphoric fumes from the manufacture of matches; and carbon disulphide from some India rubber works.

The actual weight of volatile matters thrown into the air by manufactories is often very great. The following results of analysis of flue gases of a large smelter are given by Harkins and Swain.[1]

Average in pounds per day of substances thrown off in the smoke of a large American smelter.

Arsenic trioxide	59,270
Antimony trioxide	4,320
Copper	4,340
Lead	4,775
Zinc	6,090
Oxides of iron and aluminium	17,840
Bismuth	880
Manganese	180
Silica	10,260
Sulphur trioxide	447,600
Sulphur dioxide	4,636,000

It cannot be doubted that many of the gaseous products of factories are injurious to health although it is true that, unless favored by particular conditions of wind and weather, they are not noticed by persons at more than a very short distance from the factories where they are produced. Most of these gases are well known to be injurious if present in appreciable quantity.

The following table [2] has been compiled from the reports of many investigators to show at what concentrations

[1] Journal Am. Chem. Soc., Vol. 29., No. 7, p. 970.

[2] Lehmann, Meth. Prac. Hy., 1901, p. 174.

various common industrial gases are capable of producing immediate and observable effects upon health.

Table showing concentration by volume at which certain common gases cause visible injury to health.

Name of Gas		Rapid and dangerous injury	Bearable for 30–60 min. without grave effects	Trifling symptoms after action for some hrs.
Hydrochloric acid	per 1000	1.5 to 2	.05 [1]	.01
Sulphurous acid	per 1000	0.4 to 5	.05	...
Carbonic acid	per cent	About 30	6 to 8	1 to 2
Ammonia	per 1000	2.5 to 4.5	0.3	.1
Chlorine bromine.	per 1000	.04 to .06	.004	.001
Iodine	per 1000	...	.003	.005 to .001
Hydrogen sulphide	per 1000	0.5 to 0.7	.2 to .3	.1 to .15
Carbon disulphide	per 1000	.01	.002	.001
Carbon monoxide	per 1000	2 to 3	.5 to 1.0	.2

Composition of city dusts. Beside the carbon from smoke the dust of cities is composed of particles of nearly every conceivable substance which is capable of being ground into a finely divided condition by wear and tear amid the innumerable activities of the people.

Little care is taken to prevent dust from polluting the atmosphere. Carpets and rugs are shaken on housetops, sweepings are thrown from windows, and dust composed largely of horse manure, lies in windrows on sidewalks and roadways. In nearly every large city there are factories in which large quantities of harmful dusts are produced and which are provided with blowers to force the injurious particles into the air of the streets. Building operations, overloaded coal and refuse carts, push carts and horses also contribute largely to the production of city dust.

On approaching New York from the ocean, a low, dense

[1] max. 0.1.

haze can distinctly be seen hanging over the city, and on passing in and out of the harbor a decided difference in the odor of the air may be detected. No analyses are needed to show that there are impurities in this atmosphere. The haze is largely composed of particles of soot from imperfectly consumed coal and solid matters from the factories and streets of the metropolis and cities in its vicinity.

Bacteria are vastly more numerous in the air of cities than in the atmosphere of the open country. While Miquel found, as a result of six years of observation, an average of 455 per cubic meter in the air of Montsouris Park, in the south of Paris, he found nearly 4000 in the air of the busy center of the city. The numbers found may reach almost any figure, depending upon local circumstances.

Among the worst ingredients of city dust are the parasitic bacteria. The most harmful are produced by the human body in disease and are given off in excretions of the lungs, throat, mouth and nose. Most of these bacteria die soon after leaving the body, but some resist drying and, being cast into the streets, eventually become desiccated, ground into powder and distributed through the air to be breathed with dust particles. Rarely, if ever, do the germs of common diseases grow in the dust and dirt of a city, much less originate there. The enormous numbers of harmful germs cast off by sick persons and convalescents are sufficient to make dust potentially dangerous wherever filth abounds, ventilation is inadequate or crowding excessive.

Effects of atmospheric impurities in cities. The smoke in the atmosphere of cities has been found to be injurious to plants, metal and stone. It blackens the lungs of all who breathe it.

The carbon, of which the smoke particles are chiefly composed, closes up the pores of the leaves of plants and interferes with their functions, especially the transpiration of water vapor and of air. Were it not for the fact that the breathing apparatus lies on the under side of the leaves where it is, in a measure, protected, it would be impossible for some trees to live in cities, so covered with carbon do they become. In fact the conifers, which do not change their leaves as often as do other plants, actually smother with dirt, as pointed out by Agar and others.[1]

Besides the mechanical injury produced by the soot, the sulphurous gases from coal are harmful chemically. It has been shown by several investigators that the sulphur from flue gases is highly injurious to plants. In one case reported by Haywood[2] vegetation was injured for 22 miles from a large copper smelter. In numerous instances the ground surrounding such works is entirely stripped of vegetation. Unfortunately there is no simple and exact method of detecting small quantities of sulphur dioxide in the air. By drawing air through peroxide of hydrogen, Oliver found the sulphur dioxide in London air to range between 1.5 mg. to 14.1 mg. per cubic meter of air or from 1.17 to 8.73 parts per million.

Not only have trees and other vegetation been destroyed by gases from industrial works, but cattle have been killed by feeding on forage contaminated in this way. The difficulty here was from mineral poisons. From tabulated results of analyses of fodder given by Haywood, it appears that for herbivorous animals to exist within several miles of certain American smelters, they must become confirmed arsenic eaters, with a capacity of from 3 to 10.9 grains of

[1] Journal of the Royal Sanitary Institute, Vol. 27, 1906, p. 173.

[2] Journal Am. Chem. Soc., Vol. 29, No. 7, p. 998.

arsenic per day. At autopsy there is no difficulty in discerning the injuries produced. The gastro-intestinal tract is inflamed, as is the mucous membrane of the upper air passages. The glands of the stomach and kidney in section show desquamation, cloudy swelling and fatty degeneration. The effects of these fumes upon the health of human beings must also be injurious.

Because of the sulphurous and sulphuric acids which are present in city air, the rain water of cities has a decided effect upon stone and iron. Rideal[1] found coal soot from several different cities to contain the following percentages of SO_3:

London	4.6
Manchester	4.3
Glasgow	7.9

Dust collected from 20 square yards of glass roof at the botanical gardens of Kew and Chelsea contained 5 per cent of sulphurous acid.

In New York City a peculiar yellow hue soon settles upon buildings constructed of white marble or other light-colored material, and the author's investigations show that this color is largely due to dust containing iron, acted upon by the moisture of the atmosphere. Calculations based on the consumption of horse-shoes, tires of vehicles, and the brakes, rails and machinery of street and elevated cars show that the amount of iron and steel ground up every day in the heavy and congested traffic of New York is enormous.

The effects upon human beings of breathing city dust have repeatedly been described by sanitary authorities and the principal facts with relation to this matter are

[1] Journal of the Royal Sanitary Institute, Vol. 27, 1906, pp. 76–77.

now considered to be beyond the range of speculation or hypothesis.[1]

The effects of city dust are most apparent in the respiratory apparatus, although it may produce injury to the eyes and skin. Most of the particles breathed in are caught by the mucous membranes of the throat and nose. In view of the delicacy of these structures, the burden which they are capable of withstanding seems enormous, but there is a limit beyond which even their wonderful toleration ceases. Nearly every city dweller suffers continually to a greater or less extent with chronic catarrhal or other inflammation of the nose or throat, not to mention more serious affections.

Frequency of disease due to infected air. It may be asked, if the germs of disease are so common why is it that great epidemics do not take place? As a matter of fact, great epidemics do take place, but the diseases which they carry are so common that they do not attract attention. Nor are they always mild diseases. The annual average number of deaths from diseases of the respiratory system in the registration cities of the United States was, according to the last government census, in the five years, 1900–1904, 464, while that of smallpox was 4.6.

THE AIR OF CONFINED SPACES

Composition of respired air. The difference between the principal gases in inspired and expired air is given by the following table taken from Foster.[2]

[1] Prudden — Clean Air, Medical Record, Feb. 3, 1906.

[2] Foster's Text Book of Physiology, N. Y., Edition published in 1906, pp. 440–442.

	Inspired air	Expired air
Oxygen	20.81	16.003
Nitrogen	79.15	79.589
Carbon dioxide	.04	4.38

From these figures it appears that there is little change in the nitrogen, but that the oxygen has been reduced about 4 per cent and carbon dioxide increased by nearly the same amount.

Beside carbon dioxide, expired breath contains about 5 per cent of aqueous vapor and various other gases, including slight amounts of ammonia, hydrogen and marsh gas. These are of little or no direct consequence in studies of ventilation.

The average adult takes into his lungs about 396 cubic inches or 6500 c.c. of air per minute and gives off the same amount of vitiated air which must be greatly diluted before it is generally considered suitable for further respiration. This is not saying that the carbon dioxide, itself, is dangerous to breathe even for a considerable period of time and in considerable concentration. It must be present to an extent 40 times that present when a room begins to smell stuffy and unpleasant before it produces any immediate effect and then there is merely an increase in the rate of breathing. Neither does an increase or decrease of 2 or 3 per cent in the oxygen seem to produce any evil effect. Long before the air becomes so vitiated as this, however, other impurities from the lungs make the air extremely unpleasant. Under ordinary circumstances the carbon dioxide in badly vitiated places seldom rises above 50 parts per 10,000.

The measure of dilution necessary to keep air free from odors due to the presence of people is very easily understood, for we know that while fresh air contains about .03 per cent of carbon dioxide, and air which has passed through the lungs about 4.41 per cent, the air of enclosed spaces becomes uncomfortably close and musty when the carbon dioxide exceeds .08 per cent; this is about 5 parts per 10,000 above the amount in good outside air.

It should be remembered that the odor does not depend on the amount of carbon dioxide; this gas has no odor and is only an indicator of respiratory impurity. The temperature and the amount of moisture present contribute to an important extent in making the impure character of air apparent. The higher the temperature and the greater the humidity the more noticeable the odor becomes. DeChaumont concluded that an increase of humidity of 1 per cent has as much influence on the sense of smell as a rise of 4.18 degrees of temperature.

Cause of unpleasant odors. The unpleasant odors of poorly ventilated places are due to many causes. Nearly every substance which exists possesses a characteristic odor and if no odor is noticeable in a room or other enclosed space it is because the odors are very delicate, or we have become accustomed to them, or the ventilation is sufficient to dilute them to a point beyond which they can no longer give us any sensation.

Among the most familiar odors of poorly ventilated places are those of gas lights, oil lamps, food, clothing and personal odors. The cause of the nauseous odors of human origin which commonly exist in jails, railroad cars and other places of assembly have been the subject of frequent investigation and speculation. The opinion of

Billings, Mitchell and Bergey,[1] and several others is that they are due to volatile fatty acids associated with bodily excretions and products of decomposition contained in the expired breath of persons having decayed teeth, foul mouths or certain disorders of the digestive apparatus.

It was once commonly believed that expired air contained organic matter of a toxic nature and experiments were made which seemed to show that when the fluid which can be condensed from the moisture of the breath is injected into animals, it produces death. Some doubt, however, has recently been attached to the accuracy with which these experiments were made, the present trend of opinion being that so far as gaseous, organic impurities are concerned, air can be breathed over and over again without serious physiological effects.

Amount of carbon dioxide produced by human beings. The amount of carbon dioxide produced varies with different people according to their size and activity. In their studies of the ventilation of factories and work-shops, the Committee of Parliament, already mentioned, considered that one cubic foot per hour was a probable average quantity produced by persons at work in factories.[3]

The amount of carbon dioxide produced depends upon whether one is at rest or at work, fasting or eating an abundance of food; it depends also upon the nature of the food, body weight and age. According to Landois and Rosemann,[4] expired air contains a range of from 3.3 to 5.5

[1] Smithsonian Reports, July 18, 1905, p. 389.

[2] First Report of the Departmental Committee on the Ventilation of Factories and Workshops, London, 1902, p. 94.

[3] Lehr. Physiol. des Men. Berlin, 1905, 11th Ed., p. 229.

per cent of carbon dioxide or an average of 4.38 per cent by volume.

The most accurate tests in this direction yet made in America have been those reported by Atwater and Benedict[1] in connection with calorimeter experiments at Wesleyan University. The following table shows the results which were obtained in 35 tests.

Amount of carbon dioxide exhaled by men at work and at rest at different hours of the day.

	Rest	Work
	grams	grams
$7^a - 1^p$	18.22	46.61
$1^p - 7^p$	17.89	46.47
$7^p - 1^a$	16.78	17.95
$1^a - 7^a$	10.87	11.73
Average per hour	16.11	30.71
" " "	8198 c.c.	15,628c.c.

For practical purposes we may consider that a man loads 500 c.c. of air at each breath to the extent of about 4 per cent of carbon dioxide.

Value of analyses. The usefulness of analyses depends upon the accuracy with which they are made and interpreted. They are reliable only when they are performed with suitable apparatus and by persons who have considerable skill. Up to the present, no simple methods have been devised which will enable a person not thoroughly trained to determine the chemical or bacteriological purity of air without running a large chance of obtaining very misleading results. The usual tests are valueless in inexperienced hands.

[1] Bull. 136, U. S. Dept. of Agri., 1902, p. 147.

Furthermore, unless the samples to be analyzed are collected with great care to insure that they fairly represent the whole air under consideration, serious error is practically certain to result.

Fortunately all errors made in determining the amount of carbon dioxide, the most usual test of the quality of the air, are on the side of safety and are unbalanced. The novice finds the air too impure rather than too good. The cumulative nature of the errors which may occur in these analyses undoubtedly accounts for many of the alarming reports which have been printed concerning the air of subways and tunnels based on the work of amateur investigators.

Heat produced by human beings. The amount of heat given off by the breath and skin has been variously estimated by different investigators. According to Landois and Rosemann [1] the average man produces every 24 hours per kilo of body weight 32 to 38 calories when at rest; 35 to 45 when in easy action and 50 to 70 when at hard work. If the average man is assumed to weigh 70 kilos or 154 pounds, this is equivalent to from 2240 to 4900 calories per day.

Otherwise about 1399 calories may be assumed to be given off per square meter of surface. As the average man of 154 pounds may be assumed to measure 2.09 square meters, the number of calories given off by him per day is about 2923. These estimates compare fairly well with American figures.

The observations of Atwater and Rosa [2] and Atwater and

[1] Lehr. Physiol. des Men., Berlin, 1905, 11th Ed., p. 408.

[2] Atwater and Rosa, Bull. 63, U. S. Dept. Agri.

Benedict [1] give the most accurate estimations yet made in America of the heat given off by the breath and skin. The figures are somewhat lower than those just quoted. The average of 13 experiments was 2219 calories per 24 hours for an average man at rest, and, as an average of 6 experiments, 3409 calories for a man at work.

Assuming that one-third of the complete day is spent in work and two-thirds in rest, the figures of Atwater and Benedict give 2612 calories as the quantity of heat produced by the average working man in 24 hours.

From these figures it is easy to calculate in familiar terms the heating effect of a large number of persons in a subway or other confined space. Since each calorie is equivalent to 3.968 British thermal units, 2219 calories are equivalent to 8804 B. T. U., or about two-thirds of the total heating value of one pound of good coal. If allowance is made for the unavoidable losses which occur when attempts are made to heat air by burning coal, it appears that the heat given off per day per person will be more than that produced by the use of one pound of coal. This is equivalent to about 14 cubic feet of illuminating gas.

Moisture produced by human beings. The amount of moisture given off by the breath depends somewhat upon the temperature of the surrounding air. According to Landois and Rosemann,[2] the least production of moisture occurs when the temperature is 60 degrees Fahrenheit. These authorities consider that the total quantity given off in 24 hours ranges from 330 to 640 grams. This, in familiar terms, is equivalent to about .7 of a pint to 1.3 pints of water.

[1] Atwater and Benedict, Bull. 109, U. S. Dept. Agri., p. 142, 1898–1900.

[2] Lehr. Physiol. des Men., Berlin, 1905, 11th Ed., p. 230.

Among the most accurate determinations of the moisture produced by the breath are those of Atwater and Benedict.[1] Their figures are, for periods of rest and work:

	Rest	Work
	Grams per hour	Grams per hour
$7^a - 1^p$	18.41	84.92
$1^p - 7^p$	19.28	87.29
$7^p - 1^a$	19.76	26.62
$1^a - 1^a$	17.40	20.64
Average per hour	18.70 (14 experiments)	54.87 (21 experiments)

At this rate every one hundred people at rest give to the air nearly half a gallon of water per hour.

Amount of space required for decency, comfort and safety. Authorities differ as to the amount of space required by human beings without respect to the quantity of air supplied by ventilation. A space which contains a large number of cubic feet per person provides for a storage of air which may be drawn upon in emergency, affords superior opportunities for natural air currents which aid in diffusing impurities and gives additional chances for air to enter and leave through openings to the outer air.

Billings [2] gives as the smallest amount of cubic space permissible for common lodgings and tenement houses 300 cubic feet, and for school rooms 250 cubic feet, while for hospitals he suggested 1000 to 1400 cubic feet, depending on the requirements of the particular cases to be treated. Landois and Rosemann [3] give for dwelling rooms 795 cubic

[1] Bull. 136, U. S. Dept. Agri., 1900, p. 147.
[2] Princ. of Ventilation, N. Y., 1893, p. 135.
[3] Lehr. Physiol. des Men. Berlin, 1905, 11th Ed., p. 245.

feet and for rooms occupied by the sick 994 cubic feet. For school children in England the requirement is 80 cubic feet with a floor space of 8 square feet. The standard for factories in England is 250 cubic feet per person.

No standard has ever been proposed for subways or subway cars. When the cars are full and passengers are sitting and standing in close personal contact with a roof overhead which can almost be touched by the hand, it is needless to remark that neither comfort nor decency exist. The consequences which would follow if a train of crowded cars should stop and the ventilation should in some way be entirely shut off may be calculated. In the cars of the first New York subway during rush hours, each passenger had about 2 square feet to stand on and an allowance of from 15 to 25 cubic feet of space. We may assume a reserve space of about 2000 cubic feet of air distributed between the bodies of the passengers and between their heads and the roof. If there were one hundred passengers in a car, they would give off about 850 liters of air loaded with 4 per cent of carbon dioxide per minute. Within six minutes the carbon dioxide in the air of the car would reach 36 parts per 10,000, the air would smell unpleasant, everything would be damp and the passengers would be breathing rapidly. It is practically certain that some would become nauseated and that nearly all would be panic stricken and likely to do one another bodily harm before the air became irrespirable. But if they remained quiet, they would probably not be suffocated for nearly an hour.

Fortunately it is impossible for this picture to be realized so far as ordinary operating conditions are concerned. No cars are quite tight nor could they be made so. The amount of air which could enter and leave the transoms, open doors and windows if the cars were stalled, would keep the passengers, even in subways far beneath the surface of the ground, alive indefinitely, so long as the general air of the subway remained good. It would take only a scarcely perceptible air current passing through a subway to keep the air pure enough to avoid the suffocation of the passengers.

Supply of air required. In strictness, the amount of air to be supplied per hour in order to keep the carbon dioxide of an enclosed space down to a given standard depends ultimately upon the number of persons present and the number and nature of the lights. Enclosed spaces which are continually occupied require the same amount of ventilation, irrespective of their size.

The standards which have been laid down as to the amount of carbon dioxide which ought not to be exceeded in the air of enclosed spaces have been based partly upon the unpleasant sensations produced by air and partly upon considerations of expediency: that is, what it has been found practicable to attain in the way of ventilation.

The demand for better conditions. Through the enlightening effects of education, higher standards of living are constantly being erected and these are bringing about a desire for more decent and tolerable conditions everywhere. The time has passed when sanitarians found it necessary to declare a condition perilous to health before demanding that it be improved. If air is decidedly uncomfortable by reason of odor, heat or dust, abundant warrant exists for bettering it.

In a similar way considerations of self respect are urged as reasons for demanding that adequate space be provided for people to work and live in. Those who champion this cause argue that it is degrading for human beings to be packed as close as it is physically possible for them to be packed. There is here a demand that provision be made for what may be termed decency as distinguished from health. Practically the two are inseparable.

How far the requirements of public comfort will extend in the future to underground roads it is impossible to say, but it seems not unlikely that the public will insist more and more upon having air which is clean, air which is not uncomfortably hot, reasonable freedom from unnecessary noise, ample artificial light where natural light is not procurable, abundant provision against accident and sufficient carrying capacity to enable passengers to travel with expedition and decency.

In the light of these facts, the use of steam locomotives in unventilated subways and long tunnels appears barbarous and is no longer to be thought of. It has been shown that effective systems for renewing the air are entirely practicable and by no means exorbitantly expensive if provided for when the road is designed. In the sanitary provisions of these tunnels may be found the secret of success or failure in subway ventilation as in ventilation everywhere. Good results do not come by accident. Proper ventilation can be obtained only by deliberate intention. It should be provided for in advance. The ventilation should be arranged for when the structure is designed.

Standards of purity for the air of subways. There is room for difference of opinion concerning proper standards of purity for subway air, the standards employed in the past having been generally those which hygienists have come to agree upon as suitable for workshops, schools and other public buildings.

It may be asked whether these standards are fair. In some cases they may not be sufficiently exacting, for such subways as the Metropolitan of London and the Rapid Transit of New York more nearly resemble streets than buildings. Except for the lack of sunlight, the general air of the subway may actually be better than the air of the streets. On the other hand, deep tubes such as those of London are far below the streets and every consideration makes it desirable to raise a high standard for them.

Probably a reasonable view to take is that no definite and fixed standards should be erected for all subways, but the air of each should be kept as pure as necessary to meet the sanitary requirements of the particular place in question. In other words, each subway should be considered on its own merits.

One of the earliest standards was that of Pettenkofer. Pettenkofer's limit was 10 volumes of carbon dioxide per 10,000 volumes of air, or, as he supposed, 6 volumes in excess of the proportion commonly found in the air of cities and towns. It is now known that the determinations of carbon dioxide made in Pettenkofer's time were about .5 parts too high.

It was deChaumont who proposed that the air of a room should be maintained at such a state of purity that a person coming directly from the external air should perceive no difference in odor between the room and the outside air. In order to accomplish this result, he proposed that the maximum amount of carbon dioxide admissible should be two volumes in excess of that in the outside air after assuming the latter to average about 4 parts by volume.

According to deChaumont's dictum, air ceases to be good when the carbon dioxide exceeds 8 volumes and is exceedingly bad when 10 or more volumes are present.[1]

More recently a standard proposed by Carnelley, Haldane and Anderson, for crowded schools, was 13 volumes of carbon dioxide per 10,000 volumes of air.

As a result of their work, Haldane and Osborn[2] recommended that the proportion of carbon dioxide in the air at the breathing level in factories and away from the immediate influence of special sources of contamination, such as persons or lights, should not rise during daylight, or after dark when only electric light is used, beyond 12 volumes per 10,000 volumes of air and that when gas or oil is used for lighting to not over 20 volumes after dark.

Calculation of fresh air requirements. If the amount of carbon dioxide produced per hour expressed in fractions of a cubic foot, be divided by the amount of carbon dioxide which is permissible, also expressed in this way, the quotient will be the number of cubic feet of fresh air which it is necessary to introduce in order to dilute the carbon dioxide to the proper amount. This may be represented by the following standard formula:

$$\frac{C}{P} = Q.$$

Here Q equals the number of cubic feet per hour of fresh air necessary, C the volume of carbon dioxide in cubic feet produced per hour by one person and P the amount

[1] DeChaumont, Proc. Roy. Soc. London, Vol. XXIII., p. 187.

[2] First Report of the Departmental Committee appointed to inquire into the ventilation of factories and workshops, London, 1902, p. 5.

of carbon dioxide, in volumes per 10,000 volumes of air, representing the permissible standard of respiratory impurity. It is evident that P is the difference between the amount of carbon dioxide in the outside air and in the inside air.

Now let it be assumed that the amount of carbon dioxide given off by the average adult per hour is 0.6 cubic foot and that the permissible impurity is represented by 0.0002 cubic foot.

Substituting these values in the foregoing formula, we have,

$$\frac{0.6}{0.0002} = 3000 \text{ cubic feet.}$$

This is the number of cubic feet of fresh air per hour which most sanitarians consider is necessary per person for ventilation.

The amount of fresh air supplied must vary with circumstances, however. Billings advises from 1800 for office rooms to 3600 cubic feet of fresh air per hour for hospitals.

Effects of bad ventilation. The injurious effects of bad ventilation are often considered to be due chiefly to want of oxygen, yet other factors are of much more importance.

The immediate evils are generally not due to gases, for unless gas lights are burning, there are not likely to be any injurious gases present in sufficient quantity to do harm. Finally, they cannot be the result of a deficiency of oxygen, for in nearly all cases there is sufficient oxygen present in the air of even the worst ventilated places to support life easily. In many places which appear to be badly ventilated, but in which the air has been proved by analysis to be sufficiently pure, headache and nausea have occurred.

The explanation of the trouble seems to lie in the undoubted effect which the imagination, properly excited, may have upon the sensations. A tightly closed room, hot and ill-smelling, is quite sufficient to produce discomfort.

Tuberculosis of the lungs and pneumonia are the fatal diseases most prevalent among persons living and working in poorly ventilated rooms. Both of these diseases, and in fact, practically all diseases of the respiratory tract, are caused by bacteria which gain access to the air passages. The special liability of persons who live in crowded rooms is probably due to the fact that the rooms become infested. The germs also are transmitted directly from person to person.

It seems, moreover, not improbable that an impure atmosphere, if breathed continually, may affect the vital and bactericidal powers of the cells and fluids of the upper air passages with which the bacteria come in contact and may thus predispose to infection.

For persons in perfect health, most bacteria are apparently harmless. They are caught for the most part by the moist linings of the mouth, nose and throat. and the delicate cilia of the respiratory passages, acting like countless scavengers, probably sweep them out as fast as they become entangled in the fluids with which the passages are bathed.

Most of the bacteria which are breathed in as far as the trachea and bronchial tubes are ultimately swallowed and pass into the stomach. So long as the passages remain in a healthy condition the danger appears to be comparatively slight.

The danger is greatly increased when the surfaces become injured, as, for example, in catarrh. In this case the mechanical arrangements for ejection are put out of service, the secretions are only imperfectly expelled and bacteria are sucked down into the air cells. Broncho-pneumonia and other obstructive changes are natural consequences.

Poor ventilation frequently leads to headache and, in some cases, nausea after an exposure of only a brief interval of time — too brief an interval to enable us to conclude that the harm is due to bacteria, and we must seek to account for it on other grounds. Continued work in a closely confined atmosphere reduces vigor in an unmistakable manner. It finally produces a cast of countenance which, in jails and penitentiaries, is termed prison pallor.

Inadequate ventilation in mines and in tunnel construction reduces the working capacity of both laborers and foremen to an extent which is comparable only with the effects of malarial fever. As many experienced employers have found, much more work is done, the health of all is preserved and many indirect economies result in reduced cost when a working-place is supplied with an ample volume of fresh air.

CHAPTER III

METHODS OF VENTILATING SUBWAYS

FUNDAMENTAL CONSIDERATIONS

Differences in the problems of ventilating subways, tunnels and mines. At first sight it would appear that the ventilation of city subways must be a part of a general problem which includes the ventilation of tunnels and mines, but this is not the fact. The ventilation of mines is quite a different subject from the ventilation of subterranean roads in which trains are operated. Not only are the impurities to be dealt with different but the movements of trains in subways and tunnels produce strong currents which would seriously interfere with the slow, regular movements of air which are desired in mine ventilation.

The ventilation of tunnels to be used by coal burning locomotives is also a different problem from that of ventilating city subways operated by electric power. The ventilation of steam tunnels is, in some respects, a simpler problem.

It is the gases due to the combustion of fuel which cause the air to be vitiated in ordinary railway tunnels; it is the problems which arise from the assembling of large numbers of people which must be solved in managing the air of electrically operated subways. In the one case we have to deal chiefly with poisonous gases — in the other chiefly with living agents of disease.

Necessity for skill in design and maintenance. There should be little difficulty about maintaining good sanitary conditions in any subterranean road providing the problem is handled in the right way, that is, put in the hands of persons who are skilled in sanitary matters.

It is too common to leave sanitary questions to persons whose training and experience have been exclusively in other directions and whose interests do not lead them to make a special study of these particular problems.

If the same grade of proficiency which is required of persons in charge of structural and electrical matters was demanded of those who are employed to attend to sanitary questions, the air of subways and public buildings would be better. In most cases enough work is done and enough money spent to maintain good sanitary conditions. The fault is that this work is unskillfully directed.

The correct management of the air of a subway requires a careful handling of many technical questions. The success of the work depends upon the thoroughness with which the scientific elements of the problem are taken into consideration. In fact in some modern tunnels, a highly skillful management of the air has been necessary to make construction and maintenance possible.

The difference between the conditions which surrounded the building of the St. Gothard tunnel, where 800 of the workmen died through defective hygienic arrangements, and the Simplon, where, in face of unprecedented obstacles, such a thing as vitiated air was unknown, and the hospitals empty, shows well what can be done when a serious effort is made in sanitation.

Physical principles involved. Air, like other ponderable substances, has certain physical properties which should be kept carefully in mind in even the most superficial consideration of questions of ventilation.

Air has weight and it is due to differences in the weight, caused chiefly by differences in temperature, that winds are formed and currents established which cause the atmosphere to circulate over the earth's surface, through city streets and even into houses and subways. One cubic foot of dry air weighs 536.29 grains, or 0.07661 pound, at the level of the sea and at 32 degrees Fahrenheit.

The weight of a given volume of air varies as its temperature, for it is the temperature which fixes its density. Air increases by $\frac{1}{492}$ of its volume for every rise of 1 degree Fahrenheit. Thus if the air of an enclosed space is 30 degrees warmer than the outside atmosphere, every pound of the latter will weigh about 1 ounce more than the same volume of air inside. It is for this reason that when cold air enters a room from an open window it sinks to the floor and does not mix freely unless it is deflected or agitated in some manner.

When air is heated it expands, becomes less dense, and, being then lighter than the surrounding air, rises. It is for this reason that hot air balloons go up, chimneys produce draughts, and many enclosed spaces, such as rooms and halls, are sometimes ventilated. It might be supposed that this principle could be applied to ventilate a subway, but although some exchange of air undoubtedly takes place in all subways on this account, it cannot usefully be employed to move all the air necessary.

Mechanical principles. The flow of air through large, underground passages depends not only upon its weight, but upon the frictional resistance offered by the walls of the passages. To force air through an airway by means of fans or otherwise often requires the overcoming of much frictional resistance.

The frictional resistance is proportional to the speed, to the perimeter and length of the airway and an arbitrary factor which depends upon the form and roughness of the walls. The value of this constant, termed K, varies greatly, so that it is unsafe to assume any value for it without a close knowledge of the conditions to which it is to be applied. In mine ventilation K has been assumed to vary between 0.00000000158 for a straight, brick-lined section, to 0.00000001257 for a rough and crooked one.

Books have been written upon the calculations necessary to determine the best forms and materials with which to construct ventilating apparatus. It is unnecessary to enter into these details here, but a few of the familiar formulæ will show the way in which some of the more useful calculations are made.

The following formulæ are for calculating the proper areas of passages, velocity of air, length of airways, quantities of air delivered, horse power and pressure, etc.;

Let A = area of passage in square feet,
H = horse power required for ventilation,
K = coefficient of friction due to the movement of air through the airway,
L = length of airway,
O = circumference of airway,
P = loss in pressure in pounds per square foot,
Q = cubic feet of air,
S = square feet of surface producing friction,
U = units of work in foot pounds to move the air,
V = velocity of air in lineal feet per second,
W = water gage in inches.

For the area of the cross section of the duct necessary to accommodate a given flow of air:

$$1.\quad A = \frac{KSV^2}{P} = \frac{KSV^2Q}{U} = \frac{KSV^3}{PV} = \frac{U}{PV} = \frac{Q}{V}.$$

For the horse power required to deliver the air:

$$2.\quad H = \frac{U}{33{,}000} = \frac{QP}{33{,}000} = \frac{5.2\,QW}{33{,}000}.$$

To determine the frictional resistance offered by the walls of the duct to the flow of air:

$$3.\quad K = \frac{PA}{SV^2} = \frac{U}{SV^3} = \frac{P}{SV^2 \div A} = \frac{5.2\,W}{SV^2 \div A}.$$

For the pressure in pounds per square foot:

$$4.\quad P = \frac{KSV^2}{A} = \frac{U}{Q} = 5.2\,W$$

$$= \left(\sqrt[3]{\frac{U}{KS}}\right)^2 \frac{KS}{A} = \frac{KSV^3}{Q} = \frac{U}{AV}.$$

For the quantity of air moved in cubic feet per minute:

$$5.\quad Q = VA = \frac{U}{P} = \frac{KSV^3}{P} = \sqrt{\frac{PA}{KS}}\,A = \sqrt{\frac{U}{KS}}\,A.$$

To determine the units of work in foot pounds to deliver the air:

$$6.\quad U = QP = VPA = \frac{KSV^2Q}{A} = KSV$$

$$= 5.2\,QW = 33{,}000\,H.$$

To ascertain the velocity of the air in a duct:

$$7.\quad V = \frac{U}{PA} = \frac{Q}{A} = \sqrt[3]{\frac{U}{KS}} = \sqrt[3]{\frac{QP}{KS}} = \sqrt{\frac{PA}{KS}}.$$

To determine the water gage:

$$8.\quad W = \frac{P}{5.2} = \frac{KSV^2}{5.2\,A}.$$

Physical principles involved in subway heating and cooling. The heating of subways affects the comfort of the travelling public to such an extent that it may be well briefly to refer here to some of the physical principles which must be taken into account in disposing of the heat.

The warm air of a subway is due to the heat produced by the machinery and brakes on the trains. The heat generated from the bodies of passengers is inconsiderable when compared with the amount which is due to the consumption of energy used in operating the trains. Were it not for the mechanical losses due to the production and transmission of the electric current, the heat produced in a subway would be the same as though the coal used up at the power house was consumed in fires along the track.

The only way for the heat to disappear is to escape through the walls or by the air.

When hot motors and brake shoes pass through a subway they lose their heat by radiation and conduction, the large volumes of air which flow over them greatly favoring the removal of heat in the latter manner. The heating of rooms by stoves and steam pipes occurs largely in the same way; in this case, however, the air is brought into contact with the hot surfaces chiefly through currents set up by the heat itself. The action of these currents constitutes a special form of conduction called convection.

The walls of a subway become heated through the direct effect of radiation from the trains and by absorbing heat from the air and they transmit the heat to the cooler earth beyond them. The rate of transmission, called thermal conductivity, differs with different substances. The rate varies, also, with the thickness of the body through which the heat is transmitted and the difference of temperature at the two sides.

When subways are first put in operation much of the heat produced by the cars is transmitted through the metal and masonry linings to the earth surrounding them, but in course of time this earth becomes warm also, and little more heat can be absorbed. The subway air then grows warmer and unless some special means of removing the heat is provided, the air may become very uncomfortable. The most practicable means of cooling a subway is to provide for a very large amount of ventilation.

PRACTICAL SYSTEMS IN USE

The ways in which subways have been ventilated may conveniently be considered under four separate heads:

1. By introducing or exhausting air at various points by means of fans.

2. By forcing a current of air from one end to the other of the whole line by fans.

3. By so-called natural ventilation.

4. By the piston action of trains.

The exhaust and plenum principles. Fans are almost invariably employed to exhaust air, not to supply it. They may exhaust through side chambers directly to the outside air, as in the older portions of the Boston subway, or by means of air ducts communicating at various points, as in the Severn and Mersey tunnels. In the former case a number of comparatively small ventilating fans are employed at the points where the air is to be extracted; in the latter, large central pumping plants are used. In any case fresh air is expected to enter at stations or other appropriate points as rapidly as the foul air is exhausted.

In the plenum principle the fresh air is forced in by the fans and the foul air escapes as best it can. This method is more often used to supply air during construction of deep subways than in subways after they are built.

Many arguments have been brought forward to show the advantage of renewing the air at stations rather than elsewhere.

It has been urged, for example, that the air should be exhausted between stations and allowed to flow in at the stations since (*a*) more passengers are congregated at stations than at other points and in this way they will get the freshest air; (*b*) the air in the cars is renewed at stations not between them, so the air should be at its best there; (*c*) this method would most rapidly remove smoke and heat in case of fire and give the best opportunity for escape through the stations.

Some of these arguments are valid while others involve refinements of logic which seem scarcely justified. If the air is renewed as frequently as it should be, it makes little difference from a sanitary standpoint at what places it is introduced or exhausted.

1. Action of fans applied at various points. The earliest use of a fan for assisting the ventilation of a railway tunnel is believed to have been in 1870 in connection with the Lime Street tunnel of the London and Northwestern Railroad at Liverpool.

Following the generous proportions of fans which had been employed in ventilating mines, this fan was 29½ feet in diameter and discharged its air into a conical brick chimney 54 feet in diameter at the base. The quantity of air thrown was 431,000 cubic feet per minute. The air was taken from a point midway between the two ends of the tunnel. The tunnel was 6075 feet in length.[1]

Fans in the Boston subway. The Boston subway is about 4⅓ miles long and is operated by electricity. It is used by trains and single trolley cars, most of whose routes lie in the open air. The speed is so slow that the ventilating currents set up by the moving cars are often scarcely noticeable. The typical section is 332 square feet where the subway is occupied by two tracks and 707 square feet where it is four tracks wide.

In the section of the road first built ventilating fans are placed in chambers alongside of the subway at points between stations and the air is discharged upward through grated openings in the sidewalks overhead or through short shafts to the outer air. The fans are 7 to 8 feet in diameter. They were intended to be of such capacity as to enable them to completely renew the air every ten minutes. Fig. 3.

In the section under the harbor the same general plan is followed of taking air in at the stations and removing it between stations. In this case, however, an exhaust duct has been placed along the top of the tunnel with occasional openings which can be opened or closed at pleasure. The cross section of the duct is about 48 square feet; the open-

[1] Francis Fox, Trans. Am. Soc. C. E., Vol. 54, Part C, p. 554.

ings are about 4 feet long and 1 foot 5 inches wide and they are placed at intervals of about 550 feet.

The air is withdrawn at each end of the tunnel and exhausted by means of fans through shafts about one mile apart, on the opposite shores. At the East Boston end

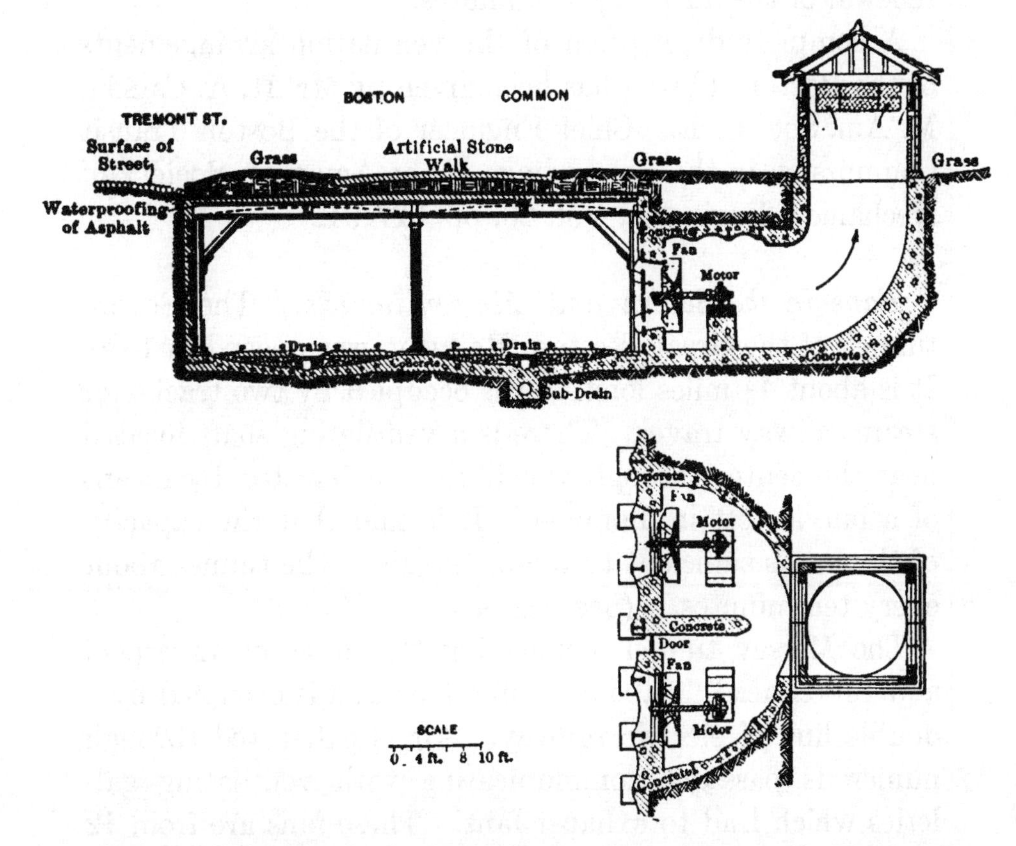

Fig. 3. Method of Ventilating the Boston Subway with Fans.

the air is exhausted through grated openings in the sidewalk 40 feet long and 7 feet 1 inch wide. At the other end the air is discharged about 21 feet above the surface of the street.

The fans consist of two 8 foot vertical fans at the East Boston end and two 7 foot horizontal fans at the Atlantic

Avenue shaft. At 175 to 218 revolutions per minute and about 12 horse power each, the total rated capacity of the whole ventilating plant is 90,000 cubic feet per minute. This gives a theoretical velocity for the whole air in the tunnel of about 2½ feet per second and is equivalent to a renewal of the air every 15 minutes.

A complete description of the ventilation arrangements of the Boston subway has been given by Mr. H. A. Carson, M. Am. Soc. C. E., Chief Engineer of the Boston Transit Commission in the Proceedings of the American Society of Mechanical Engineers, Vol. 28, pp. 927–942.

Fans in the Severn and Mersey tunnels. The Severn tunnel, of the Great Western Railway, was opened in 1886. It is about 4⅜ miles long. It is occupied by two tracks for steam railway travel. There is a ventilating shaft located near the center through which air is exhausted by means of a fan 40 feet in diameter. It is said that the capacity of the fan is sufficient to renew the air of the tunnel about every ten minutes. (See Fig. 4.)

The Mersey tunnel, connecting the cities of Liverpool and Birkenhead, is about 2 miles long and is occupied by a double line of electric railway. Air is exhausted through numerous passages communicating with ventilating galleries which lead to exhaust fans. These fans are from 12 to 40 feet in diameter and are located at stations above ground. The combined capacity of these fans is estimated to be about 950,000 cubic feet per minute or sufficient to renew the air of the tunnel every nine minutes. This tunnel is often referred to as affording an example of the most perfect system of artificial ventilation yet devised. It was certainly the earliest tunnel in which a comprehensive system was adopted. (See Fig. 5.)

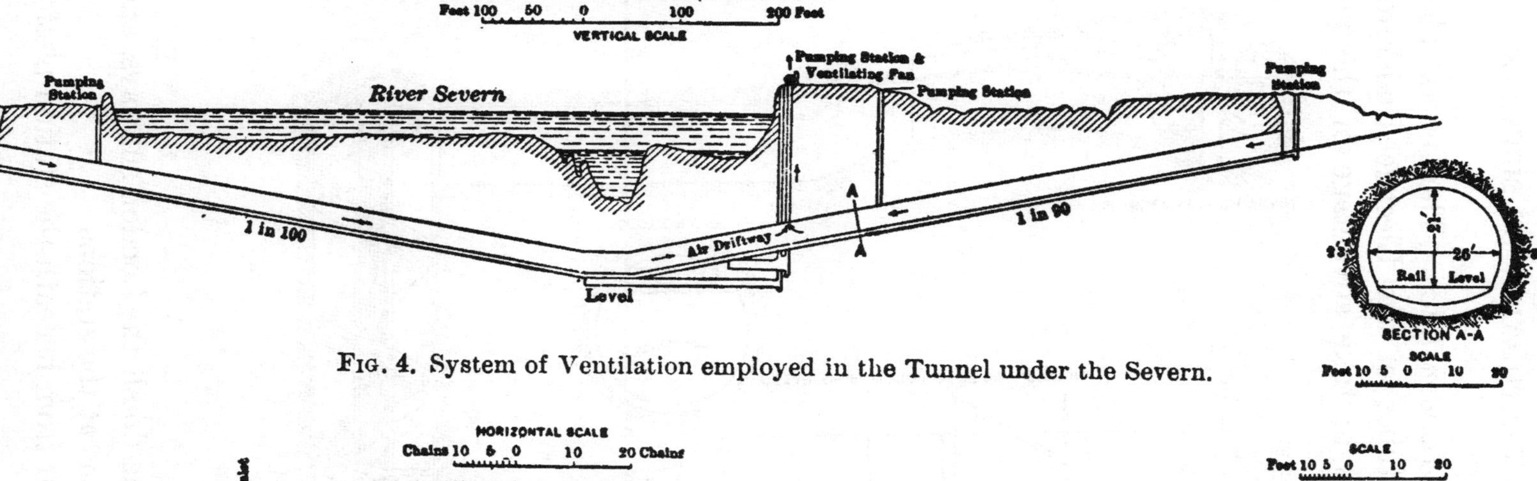

FIG. 4. System of Ventilation employed in the Tunnel under the Severn.

FIG. 5. System of Ventilation employed in the Mersey Tunnel.

Fans in the newest London tubes. The general plan of ventilating the new tubes of the Electric Underground Railways Company is to take advantage of the piston

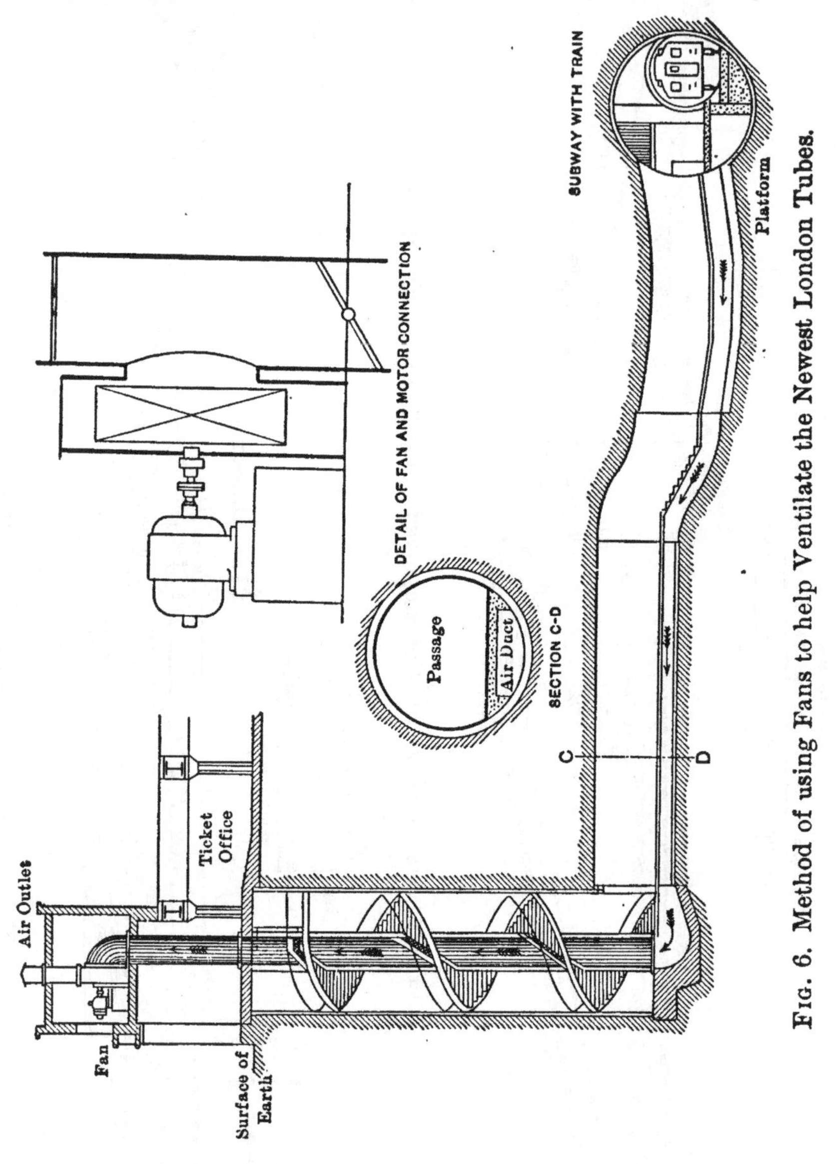

FIG. 6. Method of using Fans to help Ventilate the Newest London Tubes.

action of the trains, as do all the London subways, and to supplement this by fans at the stations.

The fans exhaust air from beneath the station platforms

and carry it through airways averaging 12 to 16 feet in cross section to the roofs of the buildings used for subway stations, there to be discharged into the free atmosphere. The fresh air enters through these station stairways and lifts. (See Fig. 6.)

The fans are of a designed capacity sufficient to remove 1,000,000 cubic feet of air per hour when working at moderate speed. This is sufficient to renew all the air in the average length of subway between two stations in each of the parallel tunnels every thirty minutes. The fans are located at the tops of the buildings. They have been found, upon test, to deliver 18,250 cubic feet per minute when operated at a velocity of 242 revolutions per minute. Great care was used to avoid vibration and noise from the motors and fans.

2. **Action of fans applied at one end of a subway.**— *System used by Central London Underground.* A system of forcing air through an electric subway has been installed in connection with the Central London Underground, a good example of the deep London tubes.

The ventilating arrangement of the Central London Railway is capable of renewing all the air contained in this subway three times over every night. In order to accomplish this result double doors are arranged at the station entrances and shut at night. The air flows in at the city end of the Bank of England, passes through the two tubes, each over six miles long, and is exhausted by a fan at Shepherd's Bush.

The fan is 20 feet in diameter, of the Guibal type. It is said to be capable of exhausting 100,000 cubic feet of air a minute as measured at a point near the far end of the line. During the day it is not possible to run this fan with much effect, because, with the opening of the station doors by passengers, it draws air from the stations, chiefly from the nearest one. But at night after the last train has been run out of the subway on the surface at the west end all the doors are closed and the fan is started. It is kept going until the first train is run in the morning. The results are said to be excellent.

3. **Natural ventilation.** Although many subways are now provided with some system of ventilation requiring the use of fans, by far the greatest number still depend for a circulation of air upon currents set up without special mechanical aid.

Blow-holes. Among the more common ways of securing this so-called natural ventilation, the use of blow-holes, or free openings to the outside air, deserve special notice. It is to ventilation accomplished in this way that the frequent renewal of air in the New York subway is due.

The draught of air passing through the blow-holes is sometimes violent. An average velocity of 16½ miles per hour through the stairways of the New York subway was observed in the author's investigations as a result of several hours' observation with anemometers. Had this current taken place through one-half of the openings between 96th Street and the Brooklyn Bridge the quantity of air so supplied would have been capable of renewing the entire atmosphere of the subway every few minutes.

At first sight it would appear that nothing could be easier than to ventilate a subway by this means. It seems as simple as opening the window of a living room. Yet to get the best effects from blow-holes, ventilation means much more than the mere opening of the roof. To provide for a suitable and reliable movement of air requires careful study. Apparently the very simplicity of the idea of blow-hole ventilation has prevented the development of this principle in the best manner. To some subways and tunnels it is peculiarly suited.

The term blow-hole is here used to include all openings through which the confined air can escape and fresh air enter, whether they be stairways, openings in the roof or openings through side chambers. In shallow subways such openings usually pierce the roof or lead from station platforms with more or less directness to the outside air. They are usually much too small, too indirect and too long to accomplish all the benefit which may be obtained from them.

Inasmuch as the flow of the air is impeded by friction against the walls, blow-holes should be as short as practicable. Since the friction increases as the square of the velocity of the current and inversely as the diameter of the passage, they should be large in section and but little obstructed by screens, doorways, nettings and other incumbrances.

It is easy to see that blow-holes may be more advantageously employed in subways built near the surface of the ground than in railways far beneath the surface. And yet this is the only way in which some of the deep London tubes are ventilated.

Direction of openings with respect to wind. If, as sometimes happens, the blow-holes are open stairways covered by cowl-like kiosks, the direction of the openings with respect to prevailing breezes may materially aid or interfere with the amount of air which passes in or out. Let us briefly examine this effect.

A breeze which is just perceptible may be assumed to travel at a velocity of 2.92 feet per second and to exert a pressure of 0.02 pounds per square foot and a breeze of twice this velocity exercises four times this pressure. A brisk wind travelling at a velocity of 25 miles per hour or 36 feet per second, a not uncommon occurrence in New York at some seasons of year, exercises a pressure of about 3 pounds per square foot. A wind of 45 miles exercises a pressure of 10 pounds. If a pleasant breeze of 2½ miles per hour acts full upon a kiosk such as many of those which stand over the New York subway, measuring 5½ × 7½ feet, it is as effective as two fans, each 6 feet in diameter, turning at the rate of 200 revolutions per minute and delivering 21,200 cubic feet of air per minute.

4. **The piston action of trains.** The action of moving trains is more important than any other factor in establishing a circulation of air through blow-holes. This so-called "piston," or "plunger" action has long been recognized as useful, but it has remained for the New York subway to demonstrate how extremely beneficial it may be.

The main principle of the phenomenon of piston action is easily understood. The moving trains force air ahead of them and cause air to rush in after they are passed.

The action is probably to be regarded as a combination of the principle of a fan in which there is practically no displacement and the principle of a plunger in which the displacement produces the whole effect. The plunger action is greatest at stations and other enlargements of the subway and where the speed is slow. The fan action is most important where the speed is high and where the train fits the subway most completely.

The quantity of air moved depends upon many circumstances. Chief of these are (*a*) the extent to which the tunnel section is filled by the section of the train; (*b*) the speed of the train; (*c*) the opportunity afforded by blow-holes for the air to flow in and out; and (*d*) the shape of the forward end of the train. These facts seem too obvious to need discussion.

Berlin-Zossen tests. In studies made on the Berlin-Zossen Railway into the resistance offered by the free outside atmosphere to the movement of trains, it was found that air piled up in front of the first car in the form of a cone of increased pressure and that a cone of reduced pressure followed behind the train. For example, the pressure in front of a car (which presented a face perpendicular to the line of the track) was 4.09 pounds per square foot at a speed of 12.4 miles per hour; 6.14 pounds at 18.6 miles; 8.19 pounds at 24.8 miles. This pressure was maintained for between 10 and 16 feet in front of the moving train; beyond this it gradually fell off.[1]

Official observations in Paris subway. A Commission appointed by the Prefect of the Seine to study the ventilation of the Metropolitan Subway of Paris gave some attention to the air currents which circulated about the trains. The air flowed ahead of the train until the front of the train was immediately opposite the observer, when there was a sudden gust and the flow changed to a direction opposite that of the train movement. After the train passed, air followed it for about one minute. When the train moved at the rate of about 3 feet per second, less than 2 miles per hour, air 165 feet ahead of the train moved at about the same velocity.

Observations in New York subway. Observations made by the author have shown that in the New York subway before any material changes were made in the arrange-

[1] Street Railway Journal, New York, October 28, 1905, p. 802.

ments for ventilation, with the ordinary train service of early afternoon, air passed from one station to another sometimes at a rate of over 8 miles per hour and at an

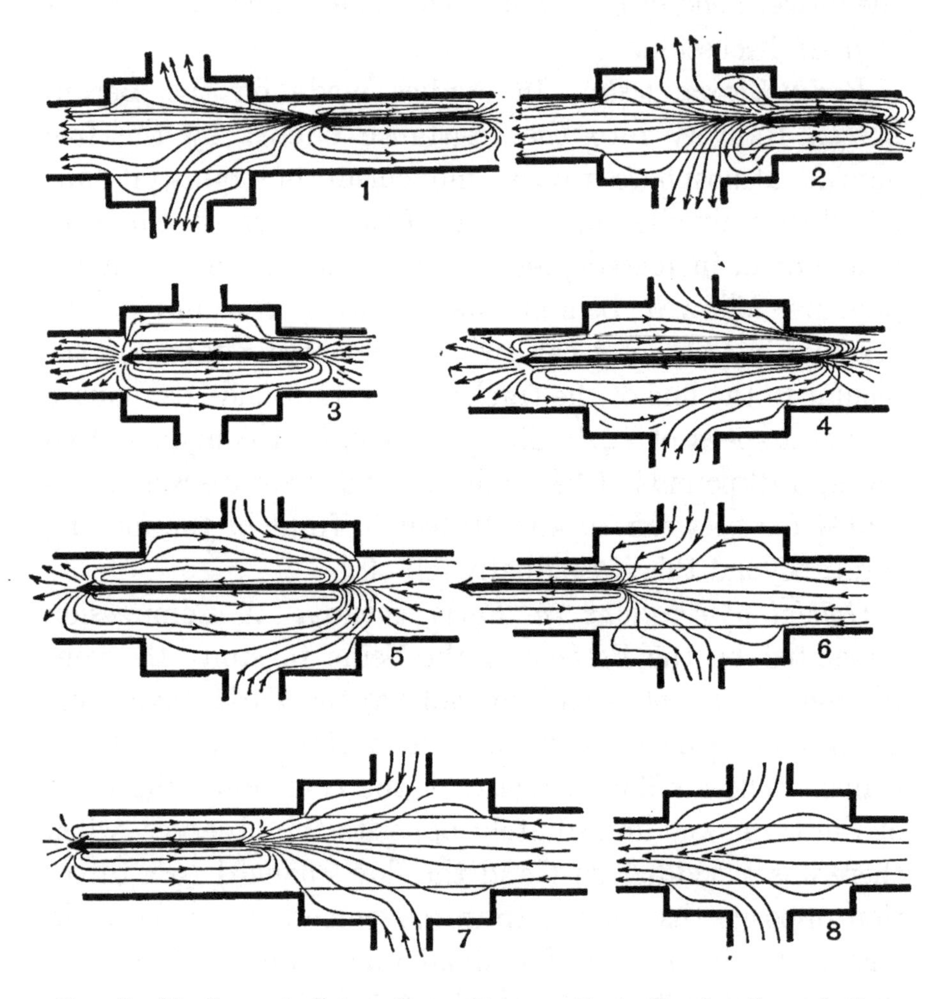

Fig. 7. Air Currents Set up by an Express Train Passing through the Simplest Form of Station in the New York Subway.

average rate of about 3 miles. The approach of a train toward a station on the four-track road could be felt by the flow of air ahead of it while the train was over 1000 feet away. (See Figs. 7, 8, 9 and 10.)

In observations made for the author in the Berlin subway, the movement of air was always easily detected while the train was 800 feet off and continued in the opposite direction for at least 25 seconds after the train had passed. In the New York subway even when no trains could be heard in either direction a distinct but faint current moved on each side of the road in the direction of the general train movement.

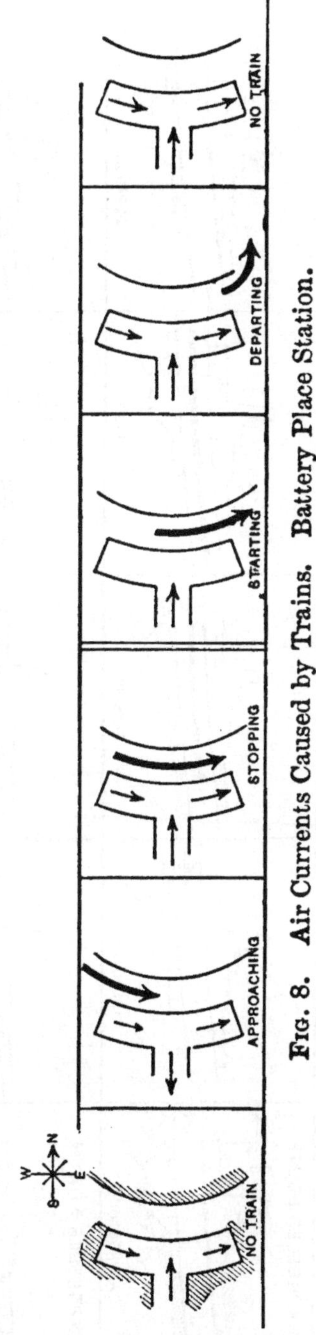

FIG. 8. Air Currents Caused by Trains. Battery Place Station.

The express trains produced the greatest amount of ventilation in the New York subway, although the action of both express and local trains was of value. The expresses were of special service in that they passed through the local stations at full speed and by their high velocity caused especially energetic currents of air to pass in and out of the stairways and openings in the roof. A somewhat detailed study was made of the direction of currents set up by express trains at several stations.

Peculiar value of piston action. The exhaling and inhaling action due to the operation of trains is of peculiar value in that it occurs when and where most needed, provided, of course, that the openings to the outside air

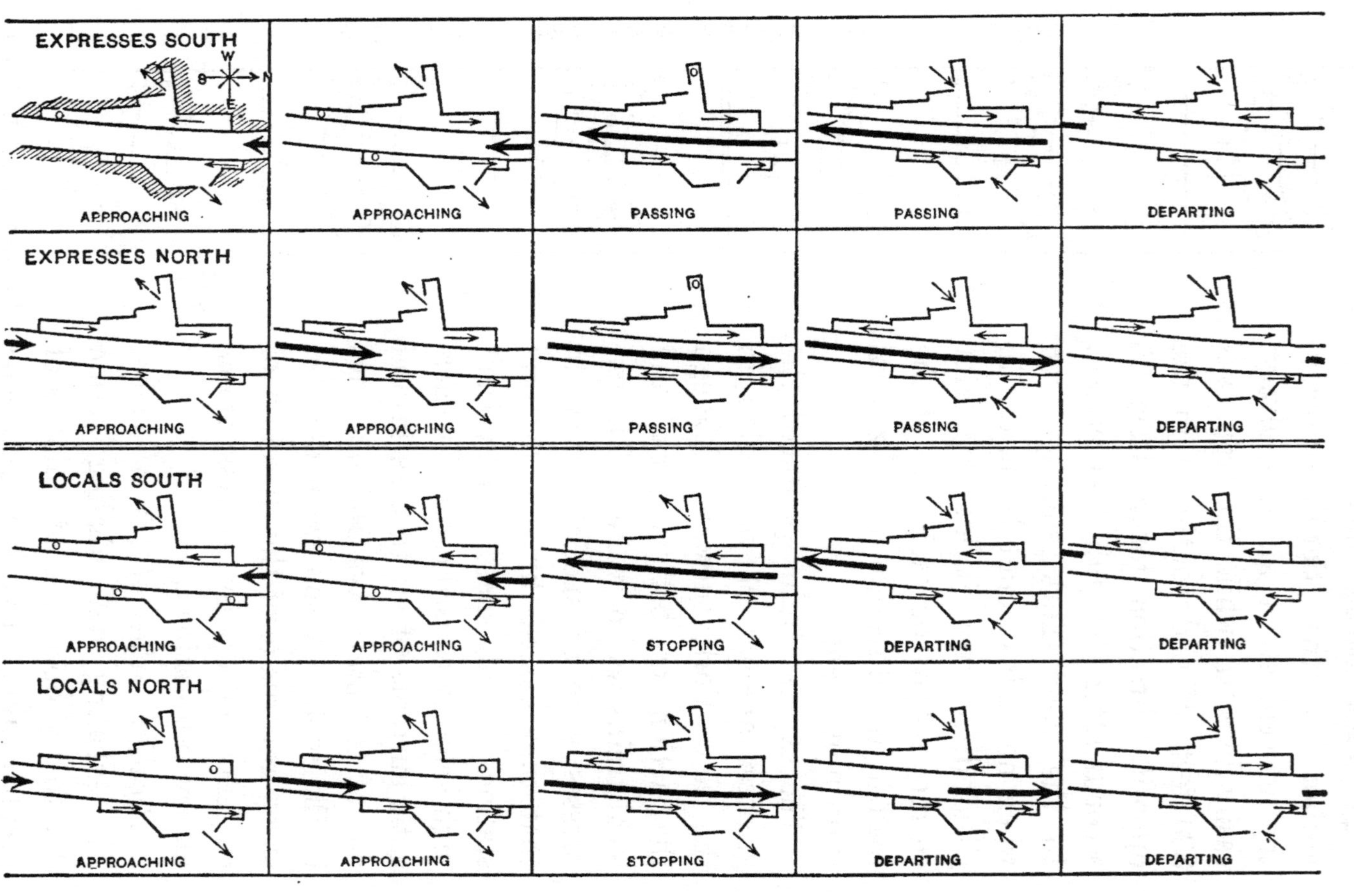

Fig. 9. Air Currents set up by Trains at Astor Place Station.

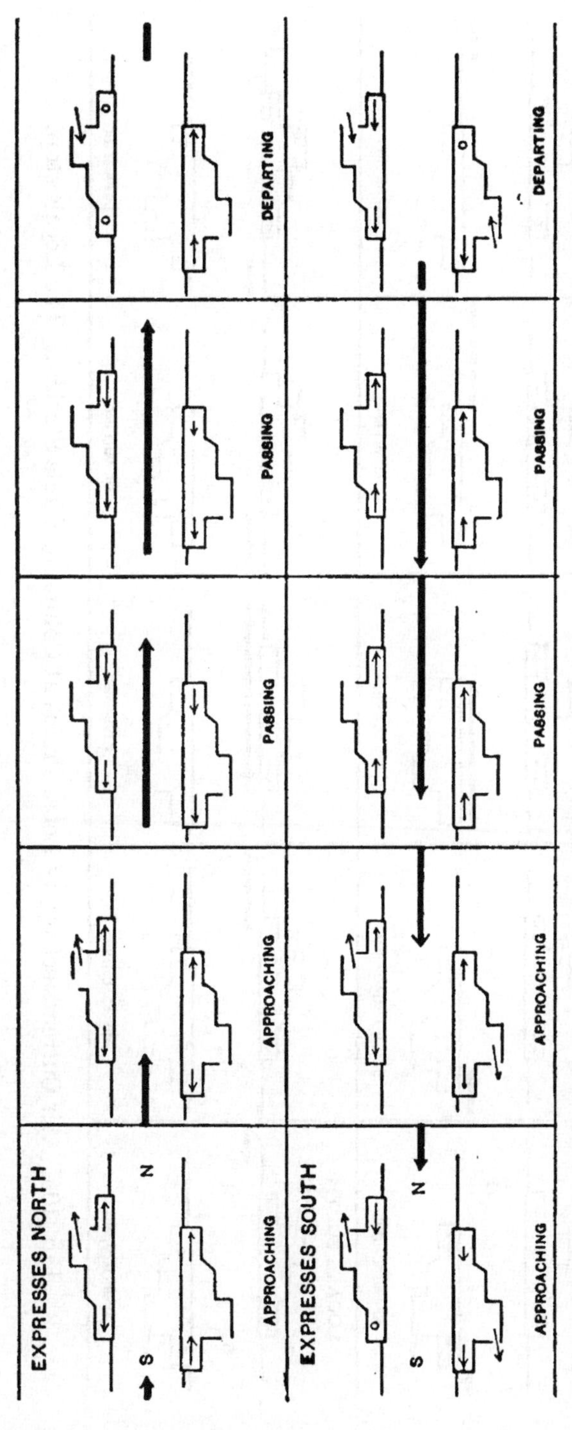

FIG. 10. Air Currents set up by Express Trains at Columbus Circle Station. The Expresses do not stop at this Station.

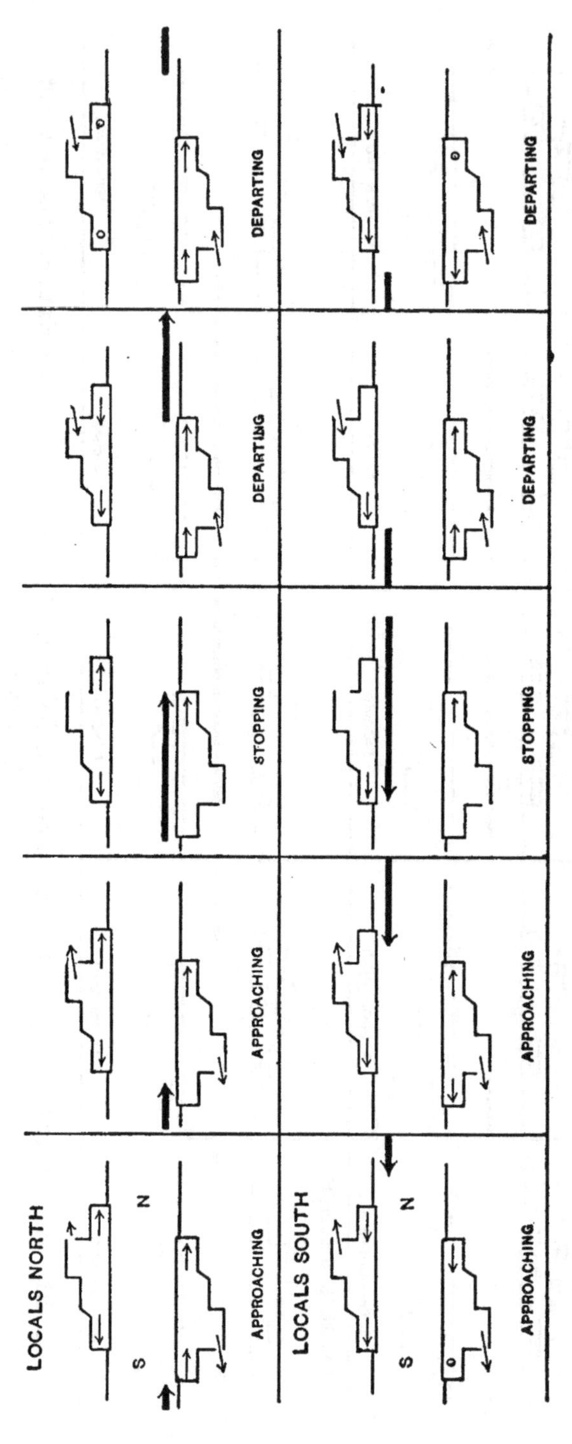

FIG. 10a. Air Currents set up by Local Trains at Columbus Circle Station. The Locals stop.

are properly placed and unencumbered. The greater the number of passengers carried and the greater the number of trains, the greater is the amount of ventilation. And not the least conspicuous of the advantages of so-called natural ventilation is its economy.

No expense is necessary for the operation of mechanical devices in natural ventilation. Experience with the New York subway shows that it is not always necessary or desirable for a train to fit very closely into the tunnel section. In fact it is conceivable that when this fit is close, the cars carry along more of their own air than desirable and the passengers within them enjoy much less interchange than would take place otherwise.

CHAPTER IV

THE AIR OF EUROPEAN SUBWAYS

METROPOLITAN AND DISTRICT OF LONDON

ONE of the first, and, in some respects, most interesting investigations yet made concerning the ventilation of a city subway was undertaken in 1897 by a Committee appointed by the London Board of Trade. The main object of the Committee was to study the system of ventilation of tunnels on the Metropolitan Railway of London and report whether any, and if so, what, steps could be taken to add to the efficiency of the ventilation in the interest of the public.

Work of an official investigating committee. The Committee consisted of Major F. A. Marindin, Earl Russell, Sir Douglas Galton, Sir Charles Scotter and Dr. John Scott Haldane. In the course of the Committee's work, thirty-six witnesses, including eminent scientists and engineers, as well as employees of the road, were called upon to testify either in the interest of the public or of the three companies which operated the road.

Among the companies' representatives were Dr. Henry Edward Armstrong, F.R.S., Sir John Wolfe Barry, Mr. Beauchamp Tower and Sir Benjamin Baker. Among other expert witnesses were Messrs. Francis Fox, Harrison Hayter, J. C. Inglis, William G. Walker and Alexander R. Binnie.

The Committee held formal meetings to take testimony and made personal inspections of the Metropolitan and other tunnels, examining air currents and taking samples of air for analysis. The final report of the Committee, containing the minutes of the meetings, testimony of witnesses and other addenda, is the source from which most of the facts concerning the investigation have been taken.

Operating conditions. During the busiest times of the day there were nineteen trains running each way per hour. In nineteen hours, 528 passenger trains and fourteen freight trains passed within the part of the line which received the greatest amount of the Committee's attention. These trains were capable of carrying 225,279 passengers. Each locomotive consumed 3 hundredweights of coal and evaporated 330 gallons of water per hour. The coal used was called "Welsh smokeless" coal. The engines were so constructed as to limit by condensation the escape of steam whilst the engine was in the tunnel.

The part of the railway which received most attention was between Edgware Road and King's Cross. The distance was about two miles. It was continuously in tunnel except at the stations. The tunnel was 16 feet high from the rail level to the crown of the arch and 28 feet 6 inches at its widest part.

The subway was ventilated chiefly by blow-holes, of which there were, in the section particularly studied, thirteen.

The management of the road admitted that the condition of air had been very unsatisfactory from the first. The original intention had been to operate the road by hot water locomotives and for this reason no structural provision had been made for ventilation except such as might occur at the staircases. The hot water locomotives proved a failure and ordinary locomotives with condensing tanks, but with ordinary coal, were at first employed. The air at once became insufferable and permission was obtained to tear up large areas of one-inch thick glass vault lights on Baker Street, Portland Road and Gower Street, and substitute gratings.

Company's experiments with ventilation. This was the beginning of a train of unsuccessful experiments which covered about 35 years. These experiments included blowing air at the rate of 60,000 cubic feet per minute into the subway through a canvas tube under the station platform at Portland Road, the idea being that the air would distribute itself under the platform and come up under the front edge so that people standing there might get the benefit of a little fresh air. This was discontinued after about 7 months, but whether because it was inefficient or because the public ceased to complain was not clear. A second fan 10 feet in diameter was connected with two brick tubes 2 feet in diameter and one-third of a mile long, opening at Baker Street, but here again the fan was worked only a short time. Blow-holes in the streets with gratings were built from time to time.

In 1871 the management changed hands and the application of the principles of fan ventilation as used in mines was considered. Sir Benjamin Baker, Past President of the Institution of Civil Engineers, after a careful study of the air currents produced by outside atmospheric conditions and by the trains, calculated that to properly ventilate the subway with fans would cost about half as much as to run the trains. The idea was to exhaust the air between stations and allow it to flow into the stations from outside.

A large number of experiments and calculations were made to determine to what extent fans could be used, but it was always shown that the continuous expense of operating fans would make them impracticable.

Following ideas successfully employed in mines considerable sums were spent in dividing the road into two parts. It was also proposed to remove the foul air produced by the locomotives by drawing it down into exhaust tubes placed between the rails. None of these schemes proved successful.

Natural ventilation. Natural currents of air passing from one station to another in this subway were observed running in velocity up to 4½ miles per hour, as measured with anemometers. The amount of air passing out of a shaft without fan ventilation was very large. In one case 171,000 cubic feet of air per minute passed out of two shafts, one 14 feet 6 inches in diameter and the other 16 feet in diameter.

Through a blow-hole about 240 feet in area at Chalton Street air flowed at the rate of from 3 to 5 feet per second. This action was due to the movement of trains and lasted from 49 to 53 seconds, an outward blast always being followed at once by an inward draught. During one hour when 31 trains passed this opening, the total period of inward draught was 34 minutes and 16 seconds and the total outward draught 26 minutes and 9 seconds.

Condition of the air. The air was described as commonly filled with smoke and steam and very unpleasant. Analyses made by Dr. Henry Armstrong, F.R.S., for the company showed that the carbonic acid sometimes amounted to as much as 65 parts per 10,000 volumes. The coal used at the time of the official investigation was analyzed by

Dr. Armstrong and described by him as an exceedingly good non-bituminous coal, containing 83.94 per cent of carbon, 8.4 per cent of ash, 1.09 per cent of total sulphur and .96 per cent of volatile sulphur. The amount of volatile sulphur which went off in the form of sulphurous acid was below one per cent.

Analyses of the air made by Dr. Haldane of the Committee gave a maximum of 89 parts of carbon dioxide. When the proportion of carbonic acid in the air did not exceed 15 parts per 10,000, Dr. Haldane considered the air "good." When present to the extent of 20 parts per 10,000, he considered it, in some cases, fairly good. Above this it was bad.

The analyses made by the Committee consisted of determinations of carbonic acid, oxygen, carbon dioxide and sulphurous acids.

The attention of the Committee was not confined to the Metropolitan Railway, but samples of air were taken for analysis from other tunnels by way of comparison. The total number of samples collected was 148.

The proportion of carbonic acid varied from a minimum of 3 to a maximum of 89. The variation was found to depend largely on the state of the wind, the movements of trains and the extent to which the subway was open to the outside air.

The oxygen was deficient in the mean proportion of 119.4 parts for every 100 parts of excess of carbonic acid (over the proportion in fresh air). In individual samples there were only slight variations from this ratio.

Carbon dioxide, due to imperfect combustion of coal, was present in very appreciable quantities. Usually there was one part of carbon dioxide to 13 parts of carbonic acid (as impurity). This proportion was fairly constant.

Sulphurous acid was present in the average ratio of one volume for every 440 of carbonic acid. The moisture on the walls of the tunnel and water standing in pools gave a marked acid reaction with litmus paper and a sample of this water gave with barium chloride a precipitate of barium sulphate showing the presence of sulphuric acid. Everything capable of corrosion in the subway appeared to be acted upon by the acid, including the leather in the boots and shoes of the employees.

The Committee, after careful consideration, were satisfied that the proportion of carbonic acid was a fair measure of the more dangerous impurities in the tunnel air and were of opinion that the maximum measure of permissible impurity should be fixed at from 15 to 20 parts of carbonic acid per 10,000 volumes of air when the outside air was normal.

Health of employees. So far as health was concerned, it appeared that no evil effects were experienced by the engine drivers, firemen and uniformed staff of employees. A list of over 1000 men all of whom had worked 5 years or more was supplied by the company. For the seven years, 1890–6, the death rate among the employees had been 11.2 per 1000, while for London it was 19.5 and for 36 towns of England 19.8. The annual average number of days' absence from sickness ranged from 7.29 in 1890 to 11.26 in 1895. The time lost through sickness was generally a little longer among drivers and firemen than among members of the uniformed force.

Methods of ventilation studied by the Committee. Two principal methods of improving the air were considered by the Committee as a result of its studies: (*a*) removal by means of fans; (*b*) additional openings.

It was recognized by the Committee that the conditions in the Metropolitan subway were quite different from those which obtained in mines where fans were successfully used. The subway was virtually a series of tunnels in short lengths. At their ends they were exposed to the action of winds. Moreover, the air was subject to the action of trains which, in the Committee's opinion, acted like ill-fitting pistons in a cylinder, continually churning up and reversing the movements of air. It was believed that these conditions would materially interfere with the effective action of fans unless the latter were made of extremely large dimensions. There was, further, the objection that the action of fans might cause enough vibration to produce a nuisance to property holders in the neighborhood of the road. Finally, it appeared certain that the fumes extracted from the tunnel would be objectionable unless discharged at a considerable height above the street pavement.

One advantage in the use of fans, if employed according to the advice of those who advocated them, was that they would improve the air at the stations where the maximum of inconvenience was felt. This would be the reverse of the conditions which obtained at first. The idea was to so place the fans between stations as to exhaust the air from each direction.

The existing system of ventilation by openings or blow-holes was considered by the Committee to be unsatisfactory both to persons using the line and to the public using the streets where the openings existed. But it appeared that a very considerable change of air was affected by large ventilating openings. Stations provided at each end with adequate blow-holes were much less uncomfortable than others.

A special point of objection was raised against the discharge of foul gases at or about the street level, both on grounds of public health and because it would deteriorate the value of neighboring property.

The Committee was convinced that pure air could best be obtained with certainty by changing the motive power to electricity, but objection to this change was made by the operating companies on the ground that they could not get a reliable firm or combination to undertake the work of electrification. It was noted, however, by the Committee at this time that the Liverpool Overhead Railway tunnel, half a mile in length, was being operated by electricity with excellent results, both as to economy and condition of the air, and the future electrification of the road was clearly anticipated.

Final conclusions of the Committee. The final conclusions of the Committee were that the most satisfactory way to deal with the ventilation of the subway would be to change the motive power to electric traction. Failing this, it would be practicable to ventilate it satisfactorily by means of fans placed at points intermediate between the stations; but the cost of doing this would be considerable. Ventilation, especially at the stations, would be sensibly improved by providing more openings or blow-holes in the roof. But satisfactory results meant a large increase in the number of openings.

In view of the fact that electric traction would probably be adopted in the near future, the Committee recommended as a temporary measure the construction of additional openings with the understanding that if electric traction was not adopted within three years, the blow-holes should be closed.

THE PARALLEL TUBES OF THE CENTRAL LONDON RAILWAY.

The Central London Railway was opened for traffic in 1900 and belongs to the type of deep tunnels, or tubes, of which London has several examples.[1]

Construction. These tubes are circular in section, lined with cast iron and lie at a depth of from 40 to 100 feet below the street level. Between stations their diameter is about 11½ feet; at the stations the diameter is about 21 feet. They are operated by electric power. Two lines of tunnel are built, each to accommodate a single track. Passengers enter the trains at the stations from platforms 300 to 400 feet long reached by elevators or, as they are called in England, lifts. It is through these passageways that ventilation is expected to take place.

Method of ventilation. Circulation of air is produced chiefly by the piston action of the trains. The cross section of the train nearly fills that of the tunnel between stations. During the busiest hours of the morning and afternoon, trains are run about every two minutes. The number of passengers carried exceeds 23,000,000 per year. The average number of passengers carried per train is 203.

Owing to what seemed to be insufficient means of ventilation due, no doubt, in part, to a consideration of the great depth of the road below the surface of the earth and to the fact that the tubes were connected with one another at frequent intervals thus interfering, to some extent, with

[1] For an account of this and other underground railways in Great Britain see Mott & Hay — Proc. International Engineering Congress, St. Louis, Mo., U. S. A., 1904, Trans. Am. Soc. C. E., Vol. 54., Part F, p. 325.

the movement of the air in and out of the passages and above all to unpleasant odors, passengers have complained that the subway was not sufficiently ventilated.

In 1901 the air of the Central London was made the subject of investigation by Dr. H. Wynter Blyth, Medical Officer for the Borough of St. Marylebone, who embodied his results in a presidential address which he delivered before the Incorporated Society of Medical Officers of Health.

Dr. Blyth's conclusion was that, so far as respiratory impurities were concerned, the air of the tube was not seriously vitiated. The analytical results showed that the general air of the Central London contained 8.7 to 10.7 volumes of carbon dioxide per 10,000 volumes of air. The outside air varied from 3.7 to 6.2.

Notwithstanding these reassuring facts, there was much popular dissatisfaction with the conditions and, owing to complaints by the public, the London County Council undertook to thoroughly investigate the condition of the air in 1902. The Council called to its assistance, as experts, Dr. Frank Clowes, Chief Chemist; Dr. Shirley Murphy, Medical Officer of Health; and Dr. Frederick W. Andrewes, Bacteriologist.

Chemical conditions. Between March and October, 1902, 118 samples of air were collected, 94 of which were analyzed chemically and 24 bacteriologically. The highest proportion of carbon dioxide was 14.7 volumes per 10,000 volumes of air; this was present in the air of a passenger carriage. The smallest, 9.6, was in an empty carriage.

About 22 per cent of the samples contained less than twice as much carbon dioxide as that found in the outside air and 34 per cent contained less than 2½ times as much as the carbon dioxide outside. Nearly all the samples seem to have been taken near noon and in only one instance apparently were more than two taken on a single day.

The air between stations ranged from 8.2 to 10.4 parts of carbon dioxide per 10,000 volumes of air. The air in the lifts contained 7.4 to 15.2 and the carriages between 9.6 and 14.7 parts. The outside atmosphere contained from 3.0 to 4.7 parts of carbon dioxide.

Samples taken on three other London underground roads during this investigation contained from 7.6 to 28.8 volumes of carbon dioxide per 10,000.

Bacteriological conditions. The bacteriological work confirmed the results obtained in the chemical analyses. Twelve samples of air from the subway were examined and compared with the results of twelve samples of air from the streets.

So far as numbers were concerned, the results corresponded closely with the conditions shown by the carbon dioxide analyses made of samples of air collected at the same time. There was a direct correspondence between the concentration of passengers and the number of microorganisms in the air.

For the fresh air of London, Dr. Andrewes in this investigation found 608 bacteria per cubic meter of air capable of growing on an agar culture medium at body temperature and ten times this number on gelatin at room temperature.

Rather more bacteria were found in the tunnel air than in fresh air samples, the average being 8820 per cubic meter, the minimum 2800 and the maximum 20,600. It is interesting to observe that the excess in numbers of bacteria in the subway over those in the street was found to be due to non-pathogenic sarcinæ and allied harmless species.

Much pains was taken to identify the species of bacteria present, but with results which were only comparatively satisfactory from a practical standpoint. The species found were in the main identical with those which occurred in fresh, outside air. No harmful kinds could be discovered. The proportion of molds to bacteria was 1 to 62.

Conclusions. The investigation led to the opinion that the air of the Central London Railway was not far different from that of inhabited rooms generally.

At the time of these studies it was proposed to flush out the air from end to end at night at the hours of minimum traffic and a system has since been adopted to accomplish this result. In the summer of 1905 automatic check valves were also in use to regulate the flow of air produced by the action of the trains.

CITY AND SOUTH LONDON RAILWAY

In 1903 an investigation of the air of the City and South London Railway was made by Dr. Scott Tebb, public analyst of Southwark.[1] This was the first deep tube subway in London and at the time of this investigation had been in operation 12 years. The carbon dioxide analyses gave the following results: In the tubes, 7.9; in the carriages in the tubes, 7.9; and in the streets, 3.8.

Bacteria were determined by exposing Petri dishes containing agar culture medium which was later incubated for 24 hours. The number of bacteria which settled from the air and were subsequently counted were: in the tunnel 57 per square foot per minute, in railway carriages 109 and

[1] Report of Public Analyst of Southwark on the Condition of the Air of the City and South London Railway, 1903, W. Scott Tebb, M.D.

in the streets 209. The fact that fewer bacteria were found in the subway than in the streets is very interesting.

THE METROPOLITAN RAILWAY OF PARIS [1]

The underground railways of the Metropolitan railway of Paris were built as the result of a law passed in 1898 which provided for an elaborate subway system to be operated by electricity. The construction has been carried on by the city. For operation the road has been leased for a term of thirty-five years to the Compagnie du Chemin de fer Metropolitain de Paris. The plan calls for eight lines of a total length of 47.74 miles. In 1907 the total length completed and in use was 27.62 miles.

The road is double tracked with a railway gauge of 4.72 feet. The width of the cars is 7.8 feet. A clearance of 1.64 feet is provided between passing cars and a clearance of 2.3 feet is required between cars and side walls. The standard section is formed by an elliptical arch having a width of 23.3 feet and a rise of 6.79 feet supported by two side walls 9.54 feet high finished inside by circular arcs. The total inside height is 17.6 feet and the width at the rail level is 21.65 feet.

A standard station comprises two side platforms 246 feet long and 13.4 feet wide. The stations are reached by staircases opening on the streets and emerging through the sidewalks without coverings. The staircases have straight flights and a width of from 9.8 to 11.4 feet; they lead downward into rooms where the tickets are sold. From the ticket rooms passengers reach the nearer platform by another staircase 9.22 feet wide, and the further platform by a similar staircase, after crossing the railway tracks by means of a foot bridge 9.8 feet wide.

[1] Biette, The Metropolitan System of Paris, Trans. Am. Soc. C. E., Vol. 54, pp. 301-324.

The masonry used is almost exclusively cement mortar and concrete. Inside the stations the walls are covered with white tiles or enameled bricks.

The tunnels of the Metropolitan system run as near as possible to the surface of the streets. In places the road emerges from below ground and follows an overhead structure, and where it has been necessary to cross the Seine Rivèr, deep tunnels or bridges have been constructed.

The subway carries large numbers of passengers and trains run frequently. In the first year 55,900,000 people were transported and in the next year over 72,000,000.

Facts concerning the provisions for ventilation have been kindly furnished by M. F. Bienvenue, Ingenieur en chef des Ponts et Chaussees, chef du Service Technique du Metropolitain. No general system either of mechanical ventilation or by blow-holes was provided, it having been believed by the French authorities that such arrangements were limited in effect to the localities where they are placed. Fans and blow-holes have been found of value, however, in a number of instances and some have been constructed and others are contemplated.

On one part of Line No. 1 which runs across the center of Paris under the rue de Rivoli and the Champs Elysee, two openings have been made which have produced much improvement. Also six similar openings have been established near the beginning of this line. Here, however, it is not the subway proper but a lateral gallery which is used for the storage of cars and to accommodate an electric sub-station that is ventilated in this way. It is proposed to open six more blow-holes to the outside air in the stations on Line No. 1.

On Line No. 2, which passes in a semi-circular way through the northern part of the city, a ventilating fan has been installed in a chamber near the western terminal of the line where the road is unusually deep. Another chamber with a ventilating shaft is installed at another point on this line and is to receive a fan also. Two large, additional openings are proposed on the boulevard des Batignolles and boulevard Belleville. At the Place de la Nation at the head of a terminal loop of the line, still another ventilating shaft exists. Finally, it is proposed to establish a second shaft a short distance from the last mentioned and still one more at the Place de la Nation.

On the southern half of this circular line the conditions of ventilation are particularly favorable. There are there only two openings for ventilation, one at the Place d'Italie over a small loop. It is proposed to make an opening at the boulevard Edgar-Quinet over a storage siding near the principal line.

On Line No. 5, a short piece of road running across the city in a generally north and south direction, it is proposed to construct six openings to the outside air; one of these is to eventually receive a fan.

On Line No. 3, the remaining section of the Metropolitan system in operation in 1907, two shafts are arranged with a view to receiving ventilating fans for the loop which exists at the terminus of this line near the Park Monceau. One other, with a fan, exists at the opposite extremity at a branch which serves for the storage of cars. Three other openings are proposed, one for each of three important stations; one of these is to be provided with a fan.

Studies of the air of the Metropolitan were begun on the 16th of January, 1901, about six months after the first section of the road, known as Line No. 1, was put in service. At a later period, Line No. 1 and the other lines as they have been successively opened have been put under systematic observation as to their temperature, humidity, and the chemical composition of the air. This work has been done by the Montsouris Municipal Laboratory. The analysts in charge have been Messrs. Albert-Levy and A. Pecoul. Frequent reports have been made by these gentlemen to the Director of the Public Works. It is to the courtesy of the gentlemen in charge of these investigations and to their official reports that the following data are chiefly due.

At least twice a year, at times corresponding to the hot and cold seasons, the investigators have visited the different parts of the lines to take samples of the air, the hour chosen usually being between 3 and 4 o'clock in the afternoon. The samples of air have been taken for analysis in rubber pouches of a capacity of about 15 liters. These samples have been transported to the laboratory for examination.

In sections of most interest, automatic apparatus has also been set up to show the chemical composition of the air during an entire day and during the day and night respectively.

In addition to the analysis of air, anemometers have been used to study the velocity of draughts and currents.

It has been found that the air improves materially during the night when no trains are operated, but when the road is put in service in the morning, the carbon dioxide increases rapidly. In the general air of the subway the carbon dioxide has been found to be generally below 9 parts per 10,000 and seldom above 16.

During the months of August and September, the proportion of carbon dioxide is notably less than at other seasons of the year, the result, no doubt, of the fact that the doors at the stations are then open and the number of passengers not as large as at other seasons. In the cars the carbon dioxide has seldom been found above 16 parts per 10,000.

In the opinion of the investigators, the discomfort which is sometimes experienced in the Metropolitan is due to the high temperature, the large amount of moisture and the amount of carbon dioxide present. In their opinion, taken alone, none of these factors would be likely to produce any discomfort, but together they produce sensations which are universally objected to.

The temperature inside of the cars does not vary greatly. It is practically like the temperature outside of the cars. In winter when the temperature of the air in the streets descends to the neighborhood of zero, it is + 20 degrees in the cars. In summer, when the temperature out of doors is 20 degrees, the cars range between 22 and 25 degrees.

The average temperature of the subway in the warm season during the years 1904–5–6 has been 20.5 degrees and 19 degrees during the cold season.

The air of the stations has been found to contain variable proportions of carbon dioxide, depending upon the composition of the air of the adjacent tunnels, the number of passengers and the frequency with which the doors at the entrances and exits are opened.

In 1903 it was noticed that black dust was soiling the walls of the tunnels and the opinion was expressed that this dust might be injurious to the employees of the road who were required to work between the stations. The dust adheres to the walls and rails. It is, moreover, regarded as inflammable as tinder. It has interfered with the insulation of the electric current.

The investigators are of opinion that the exchange of air which occurs between the subway and streets during the night is sufficient, but that that which occurs in the day is insufficient for ventilating purposes. As a result of the circulation produced by the trains, the amount of carbon dioxide increases and is distributed through the tunnels until it renders the whole atmosphere unwholesome, as measured by a standard raised by a commission on hygiene appointed by the French Minister of Commerce. This standard requires that when air contains more than 10 parts of carbon dioxide as a result of its being used for breathing, it is to be considered unsatisfactory.

It might be supposed that the circulation of the trains would produce a useful amount of ventilation, but the investigators were of opinion that this was not true. Experiments with anemometers, they thought, proved that the displacement produced by the passage of trains caused only an agitation of the air and that as soon as the train passed this agitation ceased.

The draughts of air were considered not only uncomfortable but dangerous at some of the entrances and exits of the stations. Although they seemed to some persons to indicate the displacement of a great deal of air, this displacement seemed to the investigators to be in reality very little; in fact only sudden energetic gusts and without effect upon the chemical composition of the air of the subway.

Notwithstanding these unfavorable opinions, however, the analysts have found that blow-holes do produce beneficial effects upon the composition of the air. The construction of some openings has been followed by a reduction in the carbon dioxide in the vicinity from 9.2 to 6.8 in one case, from 9.5 to 6.4 in another, from 10.3 to 6.1 in a third, and from 11.4 to 6.5 in a fourth. Nor did they find that the improvement was limited strictly to the immediate neighborhood of the openings. Otherwise it is difficult to understand why the management has decided to extend this method of relief to the extent indicated in the first part of this chapter.

Elsewhere it has been remarked that the temperature of the tunnels was generally high and did not seem to vary much with the temperature of the outside air, but it is necessary to remark that the ventilation in the evening which produces a notable reduction in the carbon dioxide also produces a slight reduction in the temperature of the tunnel. For example, when the carbon dioxide in the day has been 11.2 and the temperature 16.2 degrees, it has fallen at night with a reduction in carbon dioxide of 5.1 to 14.9 degrees.

It has also been remarked that in the early hours of the morning before the tunnel air has been at all affected by the respiration of passengers the temperature has remained high and there was a characteristic odor. The investigators think that this odor is produced by organic matter condensed upon the walls and that a very energetic ventilation would be necessary to dissipate it. If it is necessary to get rid of these odors and lower the temperature it will be necessary, in the opinion of the French experts, to employ some special process of refrigeration, such, for example, as have been used in certain mountain tunnels.

RESULTS OF AN INSPECTION OF EUROPEAN SUBWAYS IN 1907

Sensible condition of the air. In the summer of 1907 the author made careful inspections of the principal subways of Europe, including those of London, Paris and Berlin, at the request of the Interborough Rapid Transit Company of New York, and enjoyed opportunities to discuss the sanitary features of subways with the engineers who built the roads, the officials who are operating them and with chemists, bacteriologists and health officers who are interested in them from a public health standpoint.

Among those to whom the author is indebted for courtesies in this direction are Mr. William Barclay Parsons, New York; at London, Mr. Eustace Burrows, Secretary Great Northern Railway; Mr. Granville C. Cuningham, General Manager, Central London Railway; Mr. Maurice Fitzmaurice, Chief Engineer, London County Council; Mr. Frank Clowes, Chief Chemist, London County Council; Mr. James R. Chapman, General Manager, Metropolitan District Electric Traction Company, Ltd.; Mr. Francis Fox, Consulting Engineer. At Liverpool, Mr. S. B. Cottrell, Engineer and General Manager, Liverpool Overhead Railway Company. At Birkenhead, Mr. J. Shaw, Resident Engineer, Mersey Railway Company. At Paris, Mr. Georges Bechmann, Consulting Engineer, and M. Bienvenue, Chief Engineer of the Metropolitan Railway. At Berlin, Herrn A. Lerche.

There were considerable differences in the general appearance of the subways visited, and doubtless in their chemical and microbic characters, but greater differences in the sensible condition of the air. There was more or less odor about all of them, the spacious Blackwall tunnel in which no cars are operated and the elaborately ventilated Mersey tunnel not entirely excepted.

Owing, perhaps, to the hot weather, the Metropolitan of Paris was more open to the outside air than it had previously been, the doors being kept open at the stations most of the time, and the air on this road was not as unpleasant as reports had indicated.

The heat was nowhere annoying, but the air was humid and odors of disinfectants frequently produced the involuntary impression that the air was not as good as it could well have been made.

Some of the long, underground passages of the tube railways were draughty. The glazed tile linings of the stations and passageways were generally kept brightly polished and fairly clean.

Less can be said concerning the cleanliness of the floors: they were often only superficially clean. A surprisingly large amount of wood is used in the cars and on the station platforms considering the inflammability of this material and its capacity for absorbing fluid matters. The lighting of the stations and passageways was, on the whole, adequate.

Spitting on the floors was not as prevalent as might be expected in view of the fact that smoking is allowed on some roads. In London the companies recognize that they have the authority to regulate this practice and are trying to diminish it.

Some cars are well designed, well lighted and properly cleaned, but this was not the rule. In Paris the practice of leaving the management of the doors partly to the public seemed dangerous.

There was no difficulty about ventilating any of the cars. In fact the currents of air in the subways outside of the cars is so strong that with even meagre openings it was not difficult to give the passengers all the air which they will stand.

Roadbeds. When it is considered that the fate of all dirt, dust and other harmful solid matters which are not cleaned up and carried out of a subway is to get upon the roadbed and sink into it or be blown about by the trains, it would seem that the necessity for providing for thorough cleanliness in all parts would be evident. Yet in so far as the sanitary condition of roadbeds is concerned, no subway visited was wholly satisfactory. In few cases was it possible to give the whole interior the thorough cleansing which the needs of the situation demanded.

Odor. The most characteristic feature of the air of every subway was an unpleasant odor. The odor was not always the same among the subways visited and probably arose from different causes on different roads. It was interesting to observe that where two subways were connected underground, separate odors were noticeable. Doubtless these odors would have given some indication of the mixture of air which was taking place had they been studied closely.

The odors were strongest where the subways were damp and warm. The oldest roads, and particularly those which were looked after least carefully, were the most unpleasant.

In every city the odors of the subways were locally considered to be objectionable, but it was noticeable that these objections had nowhere reached such a point as to bring about a thorough investigation. A free use of disinfectants, sprinkled in liquid form, seemed to be the chief measure taken to overcome the odors. A careful study of the composition of the disinfectants and the ways in which they were used gave the impression that they were more likely to add to the difficulty than to reduce it. On one road tons of disinfectants had been used without producing any beneficial effect.

Temperature. There was considerable difference in the temperature of different subways. The shallow roads were cooler than the deep ones and the new were cooler than the old. The subways which were least open to the outer air and in which the greatest amount of travel occurred were the warmest.

It was apparent that many subways were becoming warmer year by year. The newest London tubes were from 7 to 10 degrees cooler than the older roads although the latter ran in the same clay at about the same depth and had about the same amount of travel. Observations by the managers showed that there was an increase of about 2 degrees Fahrenheit per year in the temperature of the new tubes.

Engineering opinion was unanimous concerning the cause of the heating. It is due to the heavy traffic and to the fact that the capacity of the walls and the surrounding earth to absorb heat becomes, with the passage of time, more and more nearly exhausted.

The air in the Paris Metropolitan is always warm. There is no doubt that the masonry is carrying away the heat as rapidly as its conducting power and that of the surrounding earth will permit, but there is less heat escaping through ventilation than is common in other shallow subways.

The Berlin subway is comparatively cool. The road lies, for almost its whole length, in ground water, and for this reason the heat is rapidly conducted away.

The heat of the London tubes penetrates to a considerable depth through the clay which surrounds them. Observations of temperature beyond the tubes of the Central London Railway made by Mr. Granville C. Cuningham, General Manager, showed that at a depth of 70 feet below the street level, the temperature of the earth was 65 degrees Fahrenheit, at a point 4 feet behind tunnel linings it was 63 degrees.

The air of foreign subways is comparatively dry con-

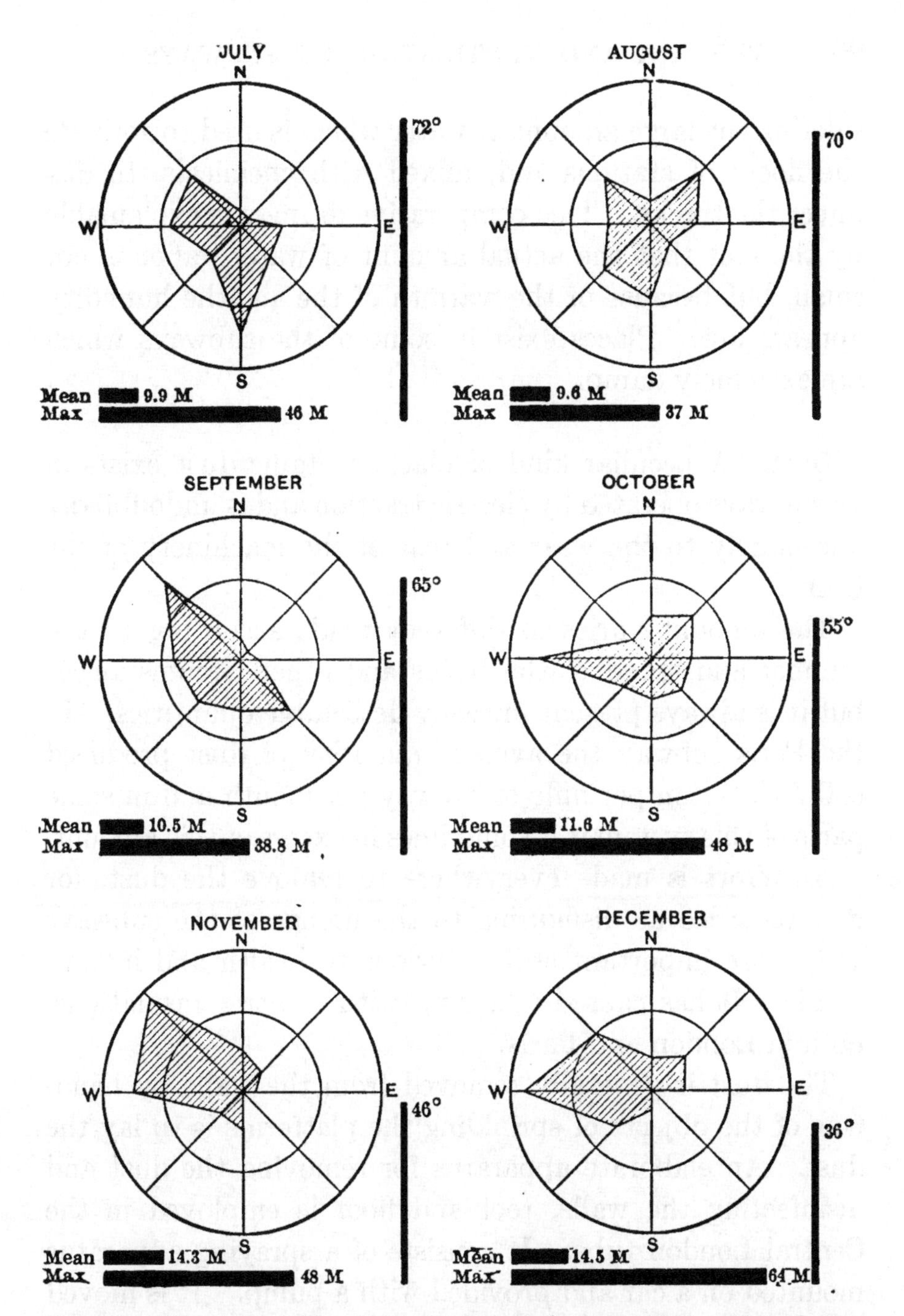

Fig. 11. Weather at New York during the investigation. Plotted from data of U. S. Weather Bureau. Shaded areas in circles are wind roses and show direction and relative movement of wind from each point of the compass. Heavy black lines below show average and greatest velocity, in miles per hour, of wind. Light black lines to the right show average temperature for the month in Fahrenheit degrees.

sidering the large amount of water which is used to sprinkle the floors of stations and, mixed with chemicals, to disinfect the tracks. This comparative dryness is explainable by the fact that the actual amount of water vapor is not small, but because of the warmth of the air, the humidity appears low. Places exist in some of the subways which are extremely damp.

Dust. A peculiar kind of black, metallic dust exists in all subways operated by electric traction and is undoubtedly due largely to the wear and tear of the machinery of the trains.

The amount varies in different roads according to the number and speed of the trains and other circumstances, but it is always present in easily detectable quantities. In the Paris subway the average quantity of dust produced is 0.7 U. S. ton per mile of subway per month and in some parts of this system it is sometimes in excess of this amount.

An effort is made everywhere to remove the dust, for it is regarded as disfiguring to the linings of the subways and, more important still, injurious to health and inflammable. It has caused difficulty with electric insulations, both in London and Paris.

The dust is generally removed from the walls by hand. One of the objects of sprinkling the platforms is to lay the dust. An elaborate apparatus for removing the dust and disinfecting the walls, roof and floor is employed in the Central London tube. It consists of a spraying apparatus mounted on a car and provided with a pump. It is moved through the subway at night and sprinkles the whole perimeter with lime disinfectant. A second apparatus formed like a large funnel is mounted on another car which runs behind and collects the dust disengaged by the sprinkler. Three of the London roads have recently introduced a new form of brake block which is said to be capable of reducing the formation of metallic dust fully 80 per cent.

Molds. Molds appear to exist in practically all European subways and are particularly abundant in damp places. One whole line, several miles in length, appears to be badly infested with these unpleasant organisms. It was said to be impossible for the management of this road to leave a car at the end of a siding for two or three days without visible growths of molds appearing over it.

It is not known that these molds are actually harmful to health, but some molds are so, and as they unquestionably produce offensive odors and are objectionable in other respects, it seems curious that no measures are taken to avoid them.

Ventilation. Owing to continued complaints by the public, methods of improving the ventilation have been adopted by various foreign subways, but nowhere has so much attention been given to this subject since electric traction came into use as in New York.

The author's studies of English, French and German subways show that the piston, or pumping, action of trains is of great value in ventilation. In fact it appears that fans and other mechanical devices are rarely necessary unless to improve purely local conditions. It was evident that the pumping action of the trains was not so complete in the deep London subways as in roads near the surface, but, combined with the action of elevators at the stations and the draughts through stairways and shafts, the trains caused an immense amount of air to pass in and out.

The subway at Berlin, which is of the same type as those

of New York, Paris and Boston, is well ventilated. The roof is close to the surface of the streets and the stairways are so arranged as to be virtual extensions of the platforms, a construction which permits of an unobstructed flow of air between the subway and the streets. Large shafts have been built to serve as emergency exits and these act as blow-holes.

Health of employees. The health of employees was inquired into without finding anywhere an excessive amount of illness which could be ascribed to subway conditions. In Paris it was thought that rather more illness, such as colds, occur among the employees, but no definite evidence to this effect could be gathered. On the London Underground Electric Railways which include the Baker Street and Waterloo, Great Northern, Piccadilly and Brompton, Charing Cross, Euston and Hempstead Railways, there are only about 0.5 per cent of the employees away through sickness at a time. Very often this figure drops to zero.

Through the courtesy of Mr. Granville C. Cuningham, the following figures were obtained to show the amount of sickness among the 500 employees at work in the Central London Railway. The following percentages refer to the month of September, 1907.

	Per cent
Inspectors, station masters and yard masters	3.3
Signalmen	0.6
Liftmen	2.5
Ticket collectors	3.0
Guards and conductors	1.0
Platform men	3.2
Average	2.3

The nature of the sicknesses is not known. Accidents were included in the figures.

CHAPTER V

THE AIR OF THE NEW YORK SUBWAY

DOUBT concerning the purity of the subway air began to occupy the public mind within a few weeks after the subway was opened in October, 1904. Some analyses, purporting to have been made in a simple and ready way by a public spirited citizen were published in the daily press and seemed to show that the air was vitiated to an alarming extent. The accuracy of these tests was challenged by competent chemists, but public anxiety, once aroused, was not easily to be set at rest.

At this point Dr. Charles F. Chandler, Professor of Chemistry at Columbia University and Consulting Hygienist to the New York City Department of Health, undertook a series of careful determinations of the carbon dioxide and oxygen in different parts of the subway and at different times of the day. By these tests the carbon dioxide was not found to be excessive nor the oxygen depleted. Dr. Chandler reached the opinion that the subway air was exceedingly good. The results of these investigations were published in the annual reports of the Rapid Transit Commission and were issued separately in pamphlet form.

The author's investigations were begun in the summer of 1905 for the Board of Rapid Transit Commissioners for the City of New York and extended continuously for six months. The original data, largely in tabulated form, were transmitted to the Commissioners in February, 1906.

It was through the intelligent and skillful efforts of

persons who were called upon to aid in the investigation that the results were in large part due. The average number of assistants continuously engaged on the work was ten. From first to last, there were twenty-one persons officially connected with the investigation; about two-thirds of this number were technical school graduates.

Capable work was done by Floyd J. Metzger, Ph.D., in the chemical analyses; by Clinton B. Knapp, M.D., and Payne B. Parsons, M.D., in the bacteriological studies; and by George S. Frost, C.E., in the meteorological observations. Aid of an unusually competent character was given in the studies of ventilation and in compiling the data by Mr. John P. Fox.

Valuable counsel and other assistance was given by a number of the author's friends who were not officially connected with the investigation. Most of these persons were professors in Columbia University. Among them may be mentioned Dr. Charles F. Chandler, Professor of Chemistry; Dr. T. Mitchell Prudden, Professor of Pathology; Dr. William Hallock, Professor of Physics; and Dr. Philip Hanson Hiss, Professor of Bacteriology. To these and others whose help added accuracy and value to the investigation, the author desires to express his sense of appreciation and thanks.

SCOPE OF THE INVESTIGATION

The principal questions investigated related to temperature, humidity, odor, bacteria, and dust. The conditions found in the subway were compared with the conditions found in the streets through which the subway runs, and occasionally with conditions in other places. The weather conditions during the investigation are shown graphically in Fig. 11.

In seeking to explain the causes of the conditions, it was necessary to take account of the sanitary care which the subway received from the company which operated it and the manner in which ventilation was accomplished.

No attempt was made to devise a comprehensive system of ventilation or cooling to improve the air. Experiments in this direction were being made by the regular engineering staff of the commission. The author merely studied the effects of these experiments and reported upon them. The details of these ventilation experiments will not be given here but it may be said that, during the operation of trains, large centrifugal exhaust fans produced no visible effect on the composition of the air, but large openings in the roof gave substantial benefit. A necessary and sufficient amount of opening for each section of the subway was clearly indicated and determined by the author's investigations.

In all, there were about 2,200 chemical analyses of air, 3,000 determinations of bacteria, and about 400 other analyses in special studies of dusts, oils, disinfectants, and other substances. About 50,000 separate determinations of temperature and humidity were made prior to the adoption of a system for automatically and continuously recording temperatures throughout the length of the subway and in the streets.

The methods employed in studying the different topics were, for the most part, such as had been used in other sanitary and meteorological investigations in which a considerable degree of accuracy was required. It is not claimed that they would have been the best to adopt in a purely scientific research. It was necessary to design them for practical as well as accurate use.

For the most part, the air to be analyzed was collected at an elevation of 18 inches to 2 feet above the pavement. This height was decided on as the most convenient and suitable, after an attempt had been made to collect it at the breathing line. Only by taking samples near the ground was it possible to avoid attracting curious crowds of persons whose presence would have rendered the samples valueless. Tests made of air from different elevations indicated that no substantial error was made in taking samples near the pavement.

Very few samples of air were taken in the cars. Persons familiar with the conditions of crowding in the cars of the New York subway at practically all hours of the day will appreciate the inconvenience with which delicate and bulky scientific apparatus could be used among persons standing as close together as it was physically possible to stand. Furthermore, the question at issue was not whether the passengers in the cars obtained good air or not, but whether the air outside the cars was satisfactory. The samples of car air analysed showed that the amount of ventilation was usually large and the air as satisfactory as could be expected considering the crowding.

ESSENTIALS OF CONSTRUCTION AND OPERATION AT THE TIME OF THE INVESTIGATION

The details of construction and equipment of the New York subway have been made the subject of so many extended and authoritative accounts that it is unnecessary to deal exhaustively with these matters here. It seems desirable, however, for the sake of clearness, to refer briefly to some of the features of construction and operation which bore directly upon the condition of the air at the time of the investigation.

Steel and concrete in the subway. The subway structure may be described as virtually a steel cage enclosed and imbedded in concrete. The walls and roof were alike in design. They found their strength in beams weighing from 42 to 70 pounds per foot, located about 5 feet apart. Between these beams square, steel rods 1¼ inches were placed to the extent of from 4 to 7 per 5-foot panel. Round rods of steel ⅝ inch in diameter connected the columns about 2 feet below the under face of the roof. The rods were set back from the inner face of the tunnel 2 inches, but the beams projected to the surface.

Between the beams of the roof and sides comparatively thin walls of concrete imbedded the steel cage. This concrete has a thickness at the walls of from 14 to 16 inches, exclusive of a thin protective wall of water-proofing outside, and of a space of variable thickness occupied by hollow ducts intended to contain electric cables. The roof has a thickness which varies from 18½ to 21½ inches.

In the four-track section of the subway especially studied in this investigation, rows of steel columns extend between each two lines of tracks at intervals of 5 feet to support the roof. (See Fig. 12.)

The floor is of concrete with an enclosed layer of water-proofing.

A report of the chief engineer of the Rapid Transit Commission supplies data from which the amount of masonry, steel and water-proofing can be calculated.[1]

Stated in round figures the quantities of material handled in building the subway were as follows:

[1] Report of Wm. Barclay Parsons, Chief Engineer, in the Annual Report of the Board of Rapid Transit Commissioners for 1904, pp. 246–251.

		Brooklyn Bridge to 50th St.	50th St. to 96th St.
Concrete and brick per mile	cu. yds.[1]	46,000	44,000
	tons[2]	93,150	89,100
Steel, including track	tons	5,300	5,400
Waterproofing	sq. yds.	82,000	124,000
Material excavated —			
Earth	per cent	76	52
Rock	per cent	24	48

Part of subway in operation. The part of the road which was in operation during the period of this investigation extended from the lower end of Manhattan Island northward to 96th Street and Broadway, where it divided, one branch continuing along Broadway to 157th Street, and the other eastward and northward until it crossed under the Harlem River and reached that part of the city known as the Bronx.

Nearly all of this road was underground. There was a short exposed portion of a few blocks covering a valley at 125th Street, and the branch to the Bronx, after crossing the Harlem, soon emerged upon an elevated structure, which it did not leave to the end of the line; but the parts of the subway which were not underground were not considered in this investigation.

Section chosen for closest observation. The considerable length of the road, about 21 miles, and the rather wide variety of conditions which occurred in it made it desirable to confine the investigation as far as practicable to a representative section.

[1] At 1 cubic foot equals 150 pounds.
[2] At 1 ton equals 2000 pounds.

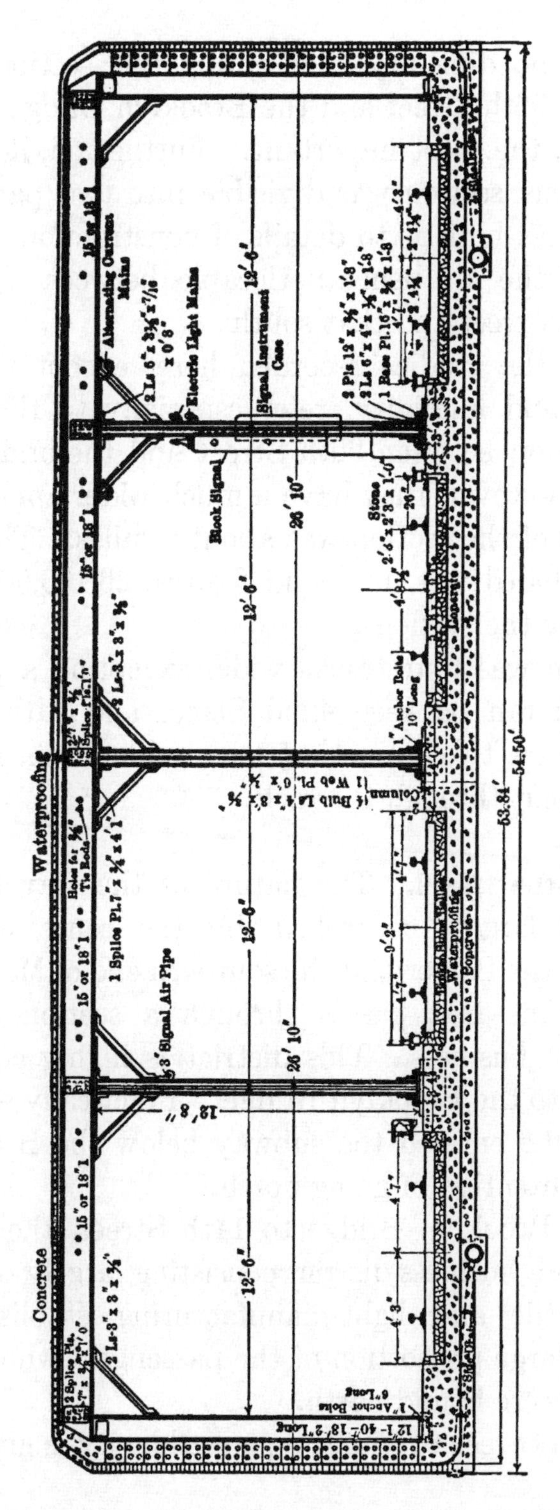

FIG. 12. Typical section of the New York Subway.

There was no difficulty in selecting this section. The road between 96th Street and the Brooklyn Bridge was, in every respect, the most important. Further on, it will be shown that this section was divisible into two parts, distinct differences both as to details of construction and the condition of the air being noticeable between the part north of 59th Street and that south.

Nearly all the studies recorded here, except those of temperature and humidity, refer especially to the representative section between 96th Street and the bridge. In many cases, however, they have a much wider application.

The length of the section was about 6 miles. The cubic air space included was, in round figures, 26,100,000 cubic feet, including the stations.

The section was four tracks wide, excepting a piece of tunnel which ran between 42nd Street and 34th Street. Here there were two tunnels of two tracks each, running side by side, cut through the rock.

Nature of the travel. The nature of the travel in the subway was largely controlled by the route followed. Beginning at the Battery, at the southern end of Manhattan Island, the line first passes through a section devoted exclusively to business. This district is highly congested and extends to the Brooklyn Bridge. Practically all of the passengers who entered the subway below the Bridge did so with the intention of going north.

From the Brooklyn Bridge to 14th Street, the subway runs through a business district consisting largely of wholesale mercantile and light-manufacturing establishments. Here also a large proportion of the passengers who entered the subway were bound north.

From 14th Street to Times Square the city is given over

chiefly to retail shops, theatres and large transient hotels. At 34th and 42nd Streets, the road is tributary to termini of two important steam railway systems. The travel in this section is heavy in each direction.

From Times Square to 96th Street, the district tributary to the subway is almost exclusively residential in character. Inasmuch as there is little to attract persons in this section to points further north, most of the passengers who enter the subway are south-bound. People who take the subway at 96th Street are also usually south-bound.

The volume of travel is indicated by the official statements of the tickets sold at different stations.[1] These statements do not mention in which direction the passengers were going, but the foregoing information concerning the districts traversed gives some idea of these facts.

Taking the sales of tickets at different stations in the month of September, 1905, we may estimate the travel as follows:

	Per cent
North-bound from Brooklyn Bridge and south	24.7
Both directions, between Brooklyn Bridge and 96th St.	44.6
South-bound, from 96th Street and north	30.7

From this it will be seen that many more people entered the subway between Brooklyn Bridge and 96th Street than passed entirely through this section on express trains.

If we divide the ticket sales in a more detailed way and apply the same reasoning as already given to show the direction of travel, we may obtain data from which to construct a diagram in which the north or south direction taken by the passengers is shown by arrows, and the comparative number of travelers, by lines of different height.

[1] Report of Board of Rapid Transit Commissioners for the City of New York, 1905.

Now, if the figures for the subway fairly represent the average conditions which occurred during the period from July to December, 1905, inclusive, we have a ready means from which to estimate the numbers of passengers carried in each direction from the different parts of the subway during this investigation. There is reason for believing that these conditions were fairly representative.

The nature of the business done by the subway and the distribution through the line indicates to some extent how the amount of travel varied from hour to hour. Between the hours of 7 and 10 A.M. and 4 and 7 P.M., the capacity of the road seemed taxed to the utmost. It is probably very close to the facts to assume that over one-half of the total travel for the day was carried within these six hours. Omitting these rush periods, there was probably a nearly constant amount of travel for the other hours of the day excepting between 12 and 6 A.M., when the amount of travel was inconsiderable in comparison with that for the other hours.

We may, therefore, without probability of serious error assume that about 50 per cent of the passengers carried each day was accommodated between 10 A.M. and 4 P.M., and 7 P.M. and 12 P.M.

If we take the total number of passengers carried on the average day in September, 1905, to have been 295,000, as shown by the official records of ticket sales, we have about 25,000 per hour as the number carried in the rush hours and 13,400 per hour during the other hours when the travel was practically at a standstill.

In the autumn of 1905, the rush-hour travel was accommodated largely by the express service. Probably 80 per cent of the travel passed entirely through the section between the Brooklyn Bridge and 96th Street.

Sanitary features of construction. By the contract we learn that it was intended, when the road was designed, that it should be easily accessible, light, dry, clean, and well ventilated.

It was partly to accomplish these ends that the road was built as close to the surface of the streets as physical conditions permitted.

Dryness. Much care was taken to make the subway dry. It was declared to be the "very essence of the specifications" for construction to secure a structure which would be entirely free from the inward percolation of ground, or outside, water.

To accomplish this end, a practically continuous sheet of asphalt was built within the concrete bottom, sides and top of the structure. This water-tight envelope consisted of from two to six thicknesses of asbestos, or other similar felt, laid in hot, natural asphalt. In addition, every part of the road was so drained that water finding access thereto was led away by natural drainage or automatic pumps to the city sewers.

Lighting. The roof of the subway was so close to the level of the streets that it was possible for the builders to make extensive use of vault lights for illuminating the stations with natural light. Full advantage was taken of the possibilities in this direction. The lights were made of cast-iron frames, with lenses about 2½ inches in diameter of strong glass set in cement. The area of the vault lights at some stations was so great that little artificial light was employed, excepting at night. Some idea of the extent to which these vault lights were used can be gained from the fact that there were over 6000 square feet at Brooklyn Bridge and over 5000 at the 96th Street station.

Incandescent lamps were the only artificial lights used except for signals.

Provisions for cleanliness. In constructing the road, provisions for keeping the subway clean were carefully carried out at the stations. The platforms were made of cement and the walls of tile, the joints and moldings being such as to permit of easy cleaning. The stairways were provided with safety treads, and these collected much street dirt, thus keeping it from entering the subway.

Provision was made in the original design for a concrete roadbed, which would have enabled the road to be kept clean between stations; but modifications in the contract, after it was let, resulted in the construction of a broken stone roadbed, from which only comparatively large particles of refuse could be removed. Smaller particles settled into the voids and finally filled them so that at many places, particularly at stations, the roadbed appeared to be perfectly smooth and black. The wooden cross-ties quickly became saturated with the oil which dripped from the machinery of the motor cars.

Stations. There were twenty stations on the section chosen for especially careful study. Five of these were for express and the remainder for local purposes only. The express stations were, on an average, 1½ miles apart and the local stations ¼ mile apart.

The express stations have two large island platforms situated between the express and local tracks. At the Brooklyn Bridge, Grand Central and 14th Street stations, these platforms are reached by overhead bridges above the tracks, but underneath the street surface. The island platforms at 72d Street are approached by stairways which descend from a more or less ornamental building located in the center of the Boulevard. The platforms of the 96th Street station are reached by a passageway running beneath the tracks. The island platforms give access to either local or express trains and permit passengers to transfer from one to another.

At the Bridge, 14th Street and 96th Street, side platforms exist in addition to the island platforms, but, excepting at 96th Street, they were not used at the time of this investigation.

The platforms of the local stations are about 200 feet long, while the platforms of the express stations are about 350 feet.

The local stations have separate platforms from which passengers enter or leave the north- or south-bound trains respectively. These platforms are located outside of the roadway. There is usually no provision for crossing from one platform to the other. At two local stations, however, Astor Place and Times Square, passageways are constructed under the railway.

There are several types of local stations. South of 50th Street there are two platforms, one on each side of the street and outside of the tracks. They are as wide as the width of the street permits. They extend from the tracks to within 5 feet of the building line. As far as practicable the two platforms are in duplicate and extend under the cross streets, sufficient street being excavated to make room for the ticket offices, toilet rooms and closets which are used for various purposes.

With few exceptions, each side of each station was equipped with two sets of water-closets for the use of women and men, respectively. They were well furnished with bowl hand basins, and sinks, each of which was supplied with a back-aired trap and connected with a sewer.

The closets were designed to be ventilated through ducts

connected with the street above, and, toward the close of this investigation, electric exhaust fans were set in many of the ducts to force the air of the closets into the street. At the beginning of this investigation, ventilation was affected only through the doors communicating with the subway.

The plans of four types of subway stations showing the stairway openings to the street are shown in Fig. 13.

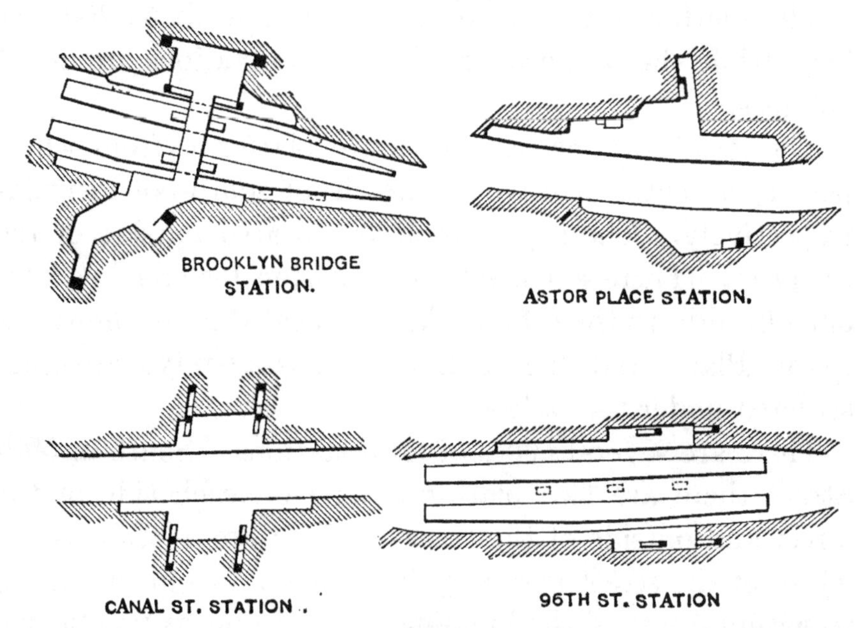

Fig. 13. Plans of four types of subway stations. Stairways and kiosk openings shown as black squares.

Track. The track consisted of a standard gauge roadway, with the wooden cross-ties and broken stone ballast, which is commonly seen on the best steam railroads out of doors. This type was adopted at the request of the operating company, after the general contract for the road had been let.

The original contract had specified that in the underground portions of the railway the tracks should consist of rails laid on a continuous bearing of wooden blocks, the grain of which was to be transverse to the length of the rail. The blocks were to be held in place by guard rails secured to metal cross-ties imbedded in the concrete floor.

Provisions for ventilation. The subway was ventilated through the stairways at the stations and through blow-holes in the roof.

Blow-holes. All of the blow-holes which were originally built were located in that portion of the road which lay above 60th Street. They were rectangular in shape, and opened upon small grass plots which occupied the center of the wide boulevard known as Upper Broadway. Iron railings surrounded the openings. To prevent the entrance of large objects, the openings were covered with coarse wire netting.

The blow-holes were located above the center of the railway, one being situated a little beyond each end of each station. An additional blow-hole was placed midway between stations. The total number of blow-holes between 59th and 96th Streets was eighteen. Each was about $7 \times 14\frac{1}{2}$ feet in the clear. Wire nettings, beams, and other objects took up about one-quarter, or more, of this space, so that the total effective area from these blow-holes was about 1,368 square feet. Early in the investigation sections of the vault lights were removed from the stations at 72d and 96th Streets and left unobstructed by nettings. The area removed at 72d Street was about 108 square feet and at 96th Street about 478 square feet. This greatly relieved the unsatisfactory condition of the air at 72d Street, where the subway entrances had been covered by a building, and at 96th Street where the roof was very low and the extent of vault lights extraordinarily great.

Stairways. The stairways between the streets and the stations varied somewhat as to width and direction. Below 59th Street they were usually placed at right angles to the line of the road; above 59th Street they were parallel to the road. There were usually two stairways, each in cross section about $5\frac{1}{2} \times 7\frac{1}{2}$ feet, to each local station above 59th Street, and eight narrower ones to the other local stations.

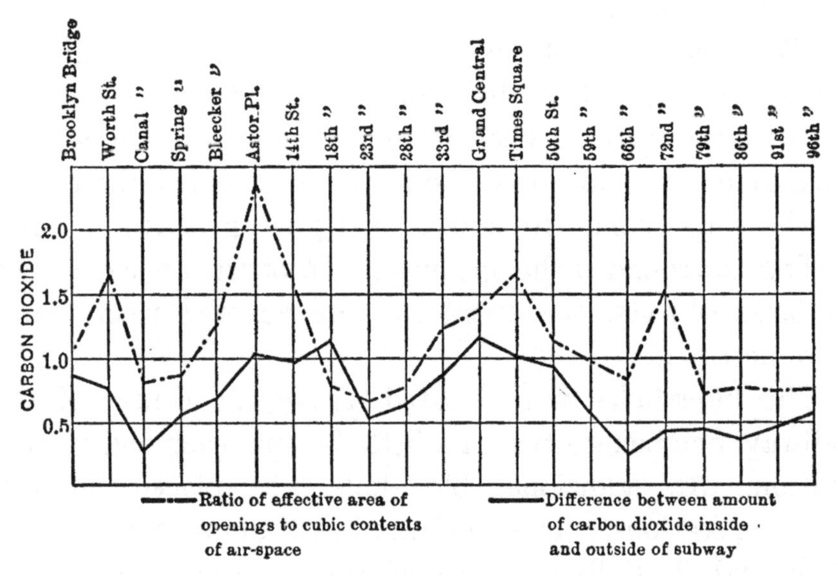

FIG. 14. Relation between the chemical condition of the air in the subway and the ratio of the effective area of the openings to the cubic contents of the air space at different stations.

The stairways were covered with ornamental kiosks, which stood upon the sidewalks near the curbs. They were without doors. The kiosks opened north and south above 59th Street; below 59th Street they usually faced east and west.

The relation between the chemical condition of the air in the subway and the ratio of the effective area of the openings to the cubic contents of the air space at different stations is shown in Fig. 14.

Phenomena of ventilation. Exchanges of air between the subway and streets took place chiefly by reason of the movement of trains.

Effects of train movements. The subway was about 50 feet wide and 18 feet high on the four-track section between Brooklyn Bridge and 96th Street, and the cross section of a car occupied about 14 per cent of this section. The trains were from about 150 feet to 408 feet long.

The number of passengers in the cars varied somewhat at different hours of the day, but the cars were usually crowded. There were fifty-two seats in each car, and when the aisles and platforms were filled the total number of passengers per car ranged from about 115 to 140. The densest crowding occurred in the rush hours and throughout the length of that portion of the subway which was under closest observation.

The number of cars per train, the number of trains per hour, and the speed varied at different hours. The local trains usually consisted of five cars, and ran at a rate, exclusive of stops, of about 21 miles per hour. The express trains generally consisted of eight cars, and ran at a rate, exclusive of stops, of about 26 miles per hour.

The total number of passengers carried in the subway, as indicated by an official statement of the ticket sales, averaged, for the last two months of 1905, 440,000 per day. There were about twice as many passengers carried in November and December as in July (see Table I).

As a train moved through the subway, air was forced ahead of it and air followed it. As a rule, a general current flowed along the track on each side of the subway in the direction of the train movement, and these currents continued even when no train was within hearing distance. The important action of a train was to force air along with it, but where stairways or blow-holes occurred and offered lines of diminished resistance, the air rushed out through them as a train approached and rushed in as the train went by.

TABLE I.

SHOWING THE NUMBER OF PASSENGERS CARRIED IN THE SUBWAY FROM JULY TO DECEMBER, 1905, INCLUSIVE.

(From the Report of the Board of Rapid Transit Commissioners for 1905.)

	Number of passengers.	Total per day.	Rush hours 7–10 A.M. 4–7 P.M.	Night 12–6 A.M.	Average.
July	6,076,241	196,000	106,000	8,000	82,000
August. . . .	7,085,814	229,000	126,000	9,000	94,000
September . .	8,843,314	295,000	159,000	12,000	124,000
October . . .	11,329,216	365,000	197,000	15,000	143,000
November . .	12,677,761	423,000	229,000	17,000	177,000
December . .	13,715,946	442,000	241,000	18,000	183,000

The difference in barometric pressure necessary to set up these air currents was exceedingly slight; the effect of friction against the walls and pillars of the subway and the sides of the stairways considerable. A great part of the force with which the air currents were set in motion was generally used up in eddies about the trains.

The movement of the air depended upon the speed of the nearest train, the movement of other trains in the vicinity, the size and location of the neighboring openings to the outside air, the size of the particular cross section of the subway with reference to the sections of the moving trains, the force and direction of the wind in the streets with reference to the position of the stairways; the difference in temperature inside and outside of the subway, and other conditions.

The chemical analyses of air which were made gave data from which the frequency with which the air was renewed could have been computed had the number of passengers present at any corresponding time and part of the subway been known. Accurate information on this subject was not, however, available. From approximate computations made in a number of ways, it is practically certain that the air was renewed at least as often as once every half hour.

TABLE II

RESULTS OF COLOGNE EXPERIMENTS, SHOWING RATE OF PASSAGE OF AIR FROM STATION TO STATION

Points of observation.	Distance traveled.	Time consumed.	Speed.
Stations.	Feet.	Minutes and seconds.	Miles per hour.
Worth Street to Brooklyn Bridge . .	835	5: 25	1.75
Worth Street to Canal Street	910	1: 15	8.27
Worth Street to Canal Street	910	7: 35	1.36
Spring Street to Canal Street	1,540	5: 30	3.18
Spring Street to Canal Street	1,795	10: 10	2.00
Spring Street to Canal Street	1,795	12: 20	1.66
Spring Street to Bleecker Street . .	1,230	3: 40	3.81
Spring Street to Bleecker Street . .	1,230	5: 25	2.58
Average	...	...	3.08

Anemometer observations. Observations with anemometers were made at a number of stations on several occasions. As a result of seventy-nine of these observations, covering, in the aggregate, two hours and thirty-five minutes, made at eight stations, it was calculated that an average of 573,000 cubic feet of air had moved in and out of one stairway per hour. This was at the rate of 9,550 cubic feet per minute.

The maximum movement of air observed was when 63,000 cubic feet passed in at one station in one minute through a single stairway. The velocity of the current on this occasion was 16⅓ miles per hour.

Circulation of air between stations. That the air circulated freely from one station to another was shown by CO_2 analyses (to be referred to later) and by noting the time that it took an odor to pass from one station to another. Cologne of a highly concentrated grade, and in sufficient quantity to produce a distinct perfume throughout the air of a station, was used at several points and the odor noted up and down the line with the help of assistants with stop watches. Care was used that the cologne should not be transported mechanically by coming in contact with the trains in liquid form. The results of the cologne experiments are given in Table II, and of CO_2 analyses in Table IV.

As a result of eight cologne experiments, it was found that the odor was carried from station to station at the

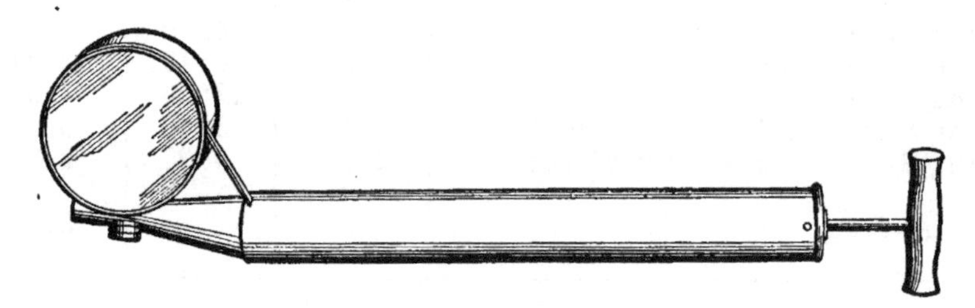

Fig. 15. Cologne Vaporizer used to determine the rate of circulation of air in the New York Subway.

average rate of 271 feet per minute, or about 3.08 miles per hour. The cologne vaporizer used in these experiments is shown in Fig. 15. It was about three feet long.

Ventilation of the subway and human lungs compared. The ventilation of the subway bears an interesting resemblance to the ventilation of the human lungs, and it will help to understand the former if we trace some of the details of this analogy.

The ventilation of both the subway and the lungs is due to currents of air passing inward and outward as a result of changes of pressure, caused chiefly by the expansion and contraction of the enclosed space.

It is true that with the lungs the size of the enclosed space is alternately enlarged and reduced through the movement of its walls, while in the subway the size of the enclosure is increased and diminished through what is termed the piston action of the trains; but in other respects the similarity is close.

In the normal amount of air which passes out of the subway on the approach of a local train, and is replaced by an indraught of fresh air as the train draws away, we have what physiologists, in speaking of the ventilation of the lungs, call the "tidal air." In the additional quantity which is drawn in by the express trains, we have the "complemental air," and in the excess which is forced out by express trains the "reserve or supplemental air."

These three, the tidal, complemental, and supplemental, we may term the "respiratory or ventilating capacity" of the subway.

Finally, there is an amount of air which remains in the subway and is not immediately forced into the streets by any combination of local and express trains; this we may call the "residual air."

This terminology is appropriate and convenient for general purposes, and it would be well if it should come into use among ventilating and sanitary experts in dealing with ventilation problems of much less strictly physiological character than those to which it has hitherto been confined.

Changes in the ventilating arrangements of the New York subway. With the object of reducing the heat which had made the air uncomfortably warm during the summer months, extensive alterations were made in the ventilating arrangements of the New York subway in 1906–7 after the investigations covered in this volume were completed. The plan embodied several features.

Large sections of the roof were removed at the stations and the openings were covered with gratings. The aggregate area of the opening, when allowance was made for the gratings, was 2,356 square feet in the section from the Brooklyn Bridge to Columbus Circle, and 1,805 square feet in the section between the latter point and 96th Street. It was calculated by the engineers of the Rapid Transit Commission that these openings, together with the openings at the station stairways, etc., would give a ratio of 1 square foot of blow-holes for every 3,200 cubic feet of contents at each station.

Blow-holes, opening generally from specially constructed chambers, were also provided between stations. These blow-holes were fitted with air valves and fans, the object of the arrangement being to induce fresh air to enter at the stations and pass out through the blow-holes between stations.

The air valves, called louvres, were made of galvanized iron and were so fitted into sheet-iron boxes that when shut they entirely closed the area of the blow-holes in which they were placed. The valves swung automatically upon axles, being so counterweighted as to open and let out air when it was forced ahead by the trains and then close and prevent any fresh air from getting in after the trains had passed.

The ventilation with the valves was like the natural

ventilation which would have taken place without them, in one respect: they were entirely dependent upon the movements of the trains, producing an amount of ventilation which was proportional to the number of trains passing in a given period.

Mr. George S. Rice, Chief Engineer of the Rapid Transit Railroad Commission, has fully described these louvres in his report to the Commission for 1906. He states that a discharge of 19,000 cubic feet of air has been observed to take place through 100 square feet of louvres between 6 A.M. and 8 P.M., an average of 14,400 cubic feet between 8 P.M. and 1 A.M. and 4,800 cubic feet between 1 A.M. and 6 P.M. under conditions of average working.

The louvre openings between Brooklyn Bridge and Columbus Circle were estimated by Mr. Rice to have a capacity of passing 675,000 cubic feet of air per minute which, being replaced by fresh air coming in through the stations, is equivalent to a renewal of all the air of this section about once in 27 minutes. Similarly the valves between Columbus Circle and 96th Street were considered to be capable of renewing the air in this section about once in 33 minutes.

The fans have been placed at the ventilating openings between stations to accelerate ventilation under special circumstances, such, for example, as at night when few trains are running and in order to free the subway of smoke in case of fire. The fans are of the centrifugal type popularly known as blowers. They are from 5 to 7 feet in diameter. They are operated by electric motors of 15 to 30 horse power capacity, and when run at their normal speed of 235 to 330 revolutions per minute are said to be capable of discharging about 990,000 cubic feet of air per minute. On the basis that the fans are really capable of operating as effectively as assumed, they should be able to renew the air in the section between Columbus Circle and Brooklyn Bridge in nineteen minutes.

A plant for cooling the air at the Brooklyn Bridge station was constructed in 1906. The project required that the heated air of this station should be passed by means of a centrifugal fan over coils of cold water and distributed through ducts opening immediately over the heads of the passengers at stations. The plant was designed by Mr. John E. Starr and is fully described in the report of Chief Engineer Rice, already referred to.

The water for the cooling plant is obtained from an artesian well from which the report says it is pumped at the rate of about 200 gallons per minute. The quantity of air cooled is about 75,000 cubic feet per minute for each of the two units into which the plant is divided.

The cooling arrangement at the Brooklyn Bridge was put in operation August 29, 1906, and has been run intermittently during the hot weather. When first put in operation, it was found that there was a transfer of heat between 9.4 and 11.3 B. T. U. per square foot per degree of difference between the air and the water, and that the air which came in contact with the pipes could be cooled about 8 degrees under the conditions of practical operation.

The flow of air from the openings in the ducts over the platforms was perceptible for a distance of 5 or 6 feet below the center of the openings. Although the regular passengers at this station took up positions immediately under the openings when the plant was operated, the effect seems not to have been perceptible to the senses at other points in the stations.

Practically none of the changes in ventilation were

completed before the investigations described in this volume were finished, although experiments with fans and blow-holes were being made toward the conclusion of the investigating work.

TEMPERATURE AND HUMIDITY

From an early period in the construction of the road, an effort had been made to observe the temperature and humidity at a number of points by means of automatic, recording thermometers. Later, when the sanitary conditions were being made the subject of investigation, these records were critically examined and the observations put upon a more exact basis.

In this section we will briefly review the early and later methods and the final system of observing temperatures, and then pass to a consideration of the results of all observations.

Methods employed. The various methods employed may conveniently be considered separately.

Early methods. The earliest date upon which any systematic record of temperature or humidity was made in the subway was June 26, 1903. At that time a number of maximum and minimum thermometers and stationary wet and dry bulb hygrometers, with some small Richard thermographs, were placed in the subway at various places between the City Hall and Columbus Circle stations, at different points on the line above 50th Street, and outside.

The accuracy of the observations made with these instruments was carefully studied by the author in July, 1905. The temperature records were found to be generally accurate, probably to within 5 degrees, and the humidity observations to within 20 per cent. The readings were all generally too high.

Methods used from July 1 *to September* 18, 1905. After the work of observing temperatures and humidities came under the author's direction, no maximum and minimum thermometers nor stationary wet and dry bulb hygrometers were employed.

Temperature and humidity observations were first undertaken in this investigation on July 1, 1905. They were made with sling psychrometers designed as made for the United States Weather Bureau, with some minor changes intended to fit them especially for subway work.

The psychrometers consisted of two mercurial thermometers 24 centimeters long, graduated from − 10 to 125 degrees Fahrenheit, fastened upon an aluminum back, 1½ centimeters apart, center to center. (See Fig. 16.)

The bulbs projected beyond the aluminum back for about 5 centimeters, one of the bulbs being covered with cotton cloth. The upper end of the aluminum back was connected by two loose wire links, fitted by a pivot to a substantial handle around which the thermometers could be whirled.

To make it conveniently transportable, the whole instrument was fitted into a specially designed cylindrical aluminum case, and was there secured by a bayonet lock at the top. The case was arranged so as to hang from a leather strap, so that the instrument might be slung over the shoulder when not in use. The psychrometers were made by Schneider Brothers, 265 Green Street, New York.

The manner in which the sling psychrometer was employed and the tables used to calculate the humidity from the readings are contained in W. B. No. 235, issued

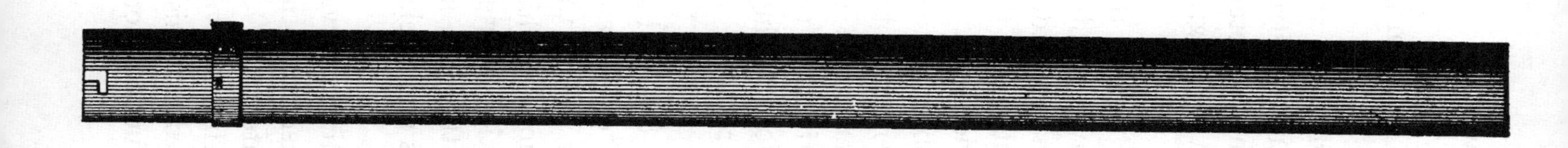

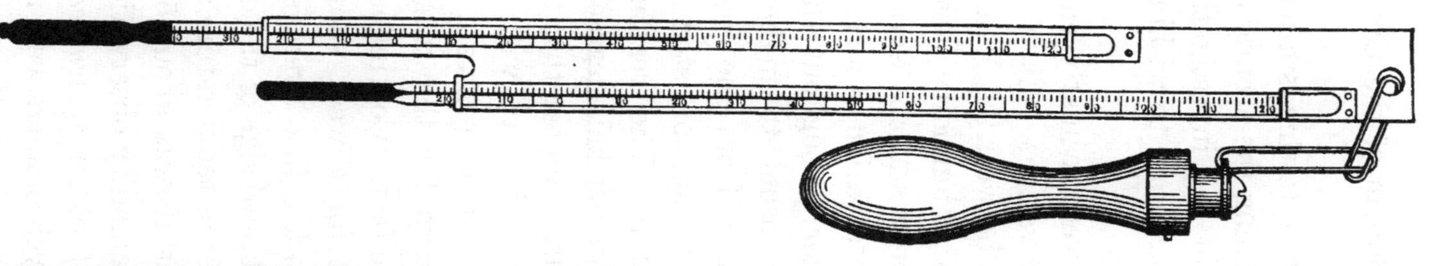

FIG. 16. Sling Psychrometer employed in determining temperature and relative humidity. Its greatest length over all was about 19 inches.

by the Weather Bureau of the United States Department of Agriculture.

When tested at a testing station, established for the purpose of determining the accuracy of the various instruments employed in the investigation, the thermometers were found to be accurate to within one-tenth of 1 degree Fahrenheit.

Four observers were regularly employed in making the psychrometer observations. Each visited a given number of stations per day, between 9 A.M. and 5 P.M., and recorded the temperature and humidity inside and outside of the subway. The express stations were visited twice.

In making an observation at a station, the observer took into account the air throughout the entire length of the platform. It was usual to make five observations at about equal distances from one end of a platform to the other. Each observation consisted of at least four readings of the thermometers. The readings were recorded in notebooks to the nearest half degree.

At first sight it may appear that these data, obtained at only certain hours of the day, with the late afternoon and early morning left out, may give a misleading idea of the average temperatures for the twenty-four hours. This, however, is not the fact. Continuous observations carried on day and night for a week showed that the data so obtained, when worked into averages for days and weeks, gave an excellent knowledge of the conditions.

The total number of temperature and humidity observations to November 16 was about 50,000, considering the temperature and humidity observations separately.

Final system of observing temperatures. The observations with sling psychrometers were not intended to furnish the data from which the most important deductions concerning the temperature were to be derived. The instruments selected for this purpose were thermographs.

The suitability of practically every well-known thermograph on the market was considered for this use. Specimens of five makes were set up in the testing station and compared for several weeks before a choice was finally made.

The instruments chosen were the largest thermographs made by J. Richard, Paris. About one dozen of these were especially imported for subway use; others were obtained

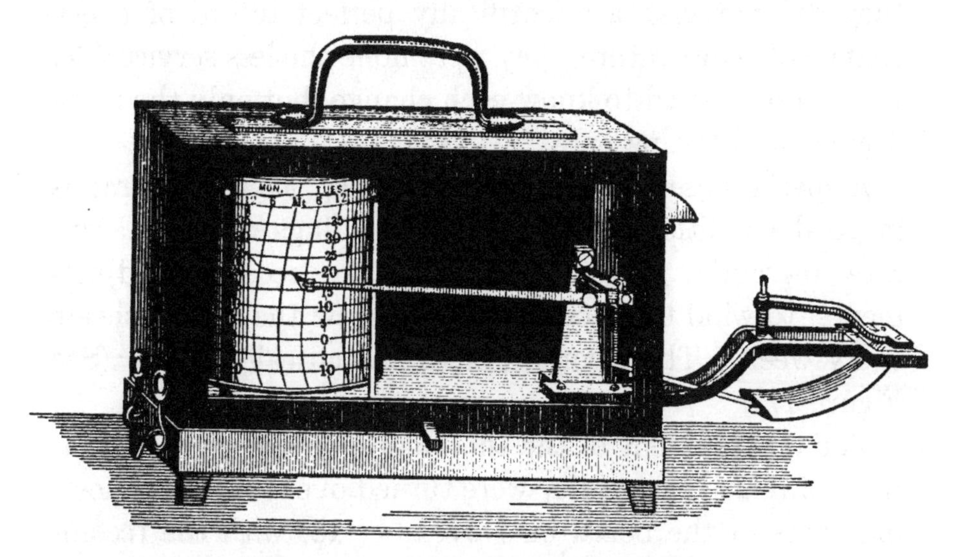

Fig. 17. A Richard Thermograph.

in America. The total number of thermographs which were put in use was twenty. On test they were found to be capable, under the practical conditions required, of recording changes of temperature to within one degree and time to within ten minutes, if visited daily and kept clean and adjusted.

As may be supposed, the thermographs never showed the changes of temperature which actually occurred. They were most accurate when the changes were slow and not very decided. To be precise, the thermograph curves could be interpreted as neither more nor less than the record of the net effect of several conditions, only the leading one of which was the temperature of the subway near the instrument.

Observations with a delicate mercurial thermometer showed that changes of temperature occurred in the subway of far greater extent and frequency than were indicated by the thermographs. (See Figure 4.) But if they did not give a scientifically perfect record of every change of temperature, they were none the less serviceable. It was not desired to know each change, but only the more decided and lasting ones.

A special observer was trained to keep the thermographs in good condition, to check their readings with an accurate mercurial thermometer, to make necessary adjustments, to wind the clocks, and to renew the record sheets every week. The records were systematically filed. (See Figure 17.)

The management of the thermographs and the observations of relative humidity were turned over to the engineering corps of the board on November 18, with the recommendation that the system be made permanent.

The thermographs were placed in specially constructed cages in the stations of the subway, generally at the end of the platform which was approached by incoming trains. They were as far removed as possible from local draughts, and about 4 feet above the pavement.

The thermographs located out of doors were placed in cages of similar construction to those used underground. The locations of the outside thermographs were selected with a view to having the instruments in the shade, remote from the effects of radiation from pavements and buildings, and beyond the influence of outrushing currents of air from the subway.

The records obtained from the thermographs gave an excellent idea of the average differences in temperature which existed between the subway and streets and between one subway station and another. They also showed fluctuations in temperature which gave clews to the origin of the heat, to the extent and ways in which ventilation took place, to the movements of trains, and many other instructive matters.

Throughout the period of this investigation, observations of temperature were made by the company which operates the subway. These observations consisted of hourly readings of ordinary tin-backed thermometers inside and outside of the stations. The thermometers were usually read by colored porters, few of whom were competent to do this work properly. None of these records is used in this paper.

Results of temperature and humidity observations.— *Before the subway was opened for travel.* In view of the overheating of the subway, which has been a characteristic of the road since 1905, it is interesting to review some of the temperature characteristics of the road before it was thrown open to travel. In reviewing these figures, it must be remembered that the entrances, exits, and blow-holes were more or less closed at this time, and that a free circulation of air in and out was correspondingly obstructed. Had all the openings been free, there would have been less difference between the inside and outside air.

During construction, and before the subway was covered

by a roof, its temperature was practically that of the outside air.

After the roof was put on, marked differences occurred. The temperature of the subway was now much more constant than that outside.

Before trains were run, hourly and daily changes of temperature in the streets seemed to affect the subway but little. The more marked and continued changes, however, produced visible effects.

In winter the air outside was cooler, and in summer warmer, than the air of the subway. The total range of temperature of the subway air at 112th Street for the year, August 6, 1903, to July 27, 1904, was from 22 to 68 degrees, or 46 degrees; while the range outside was from 2 to 94 degrees, or 92 degrees. In other words, the range of temperature in the streets was twice as great as the range in the subway.

During the hottest week in the subway, before it was opened, the temperature, according to the 112th Street records, averaged 59.4 degrees, while the street temperature was 77.7 degrees. At its hottest, the subway was 18.3 degrees cooler than the streets.

The early records of temperatures in the subway give no idea of the normal temperature of the earth through which the subway runs. The temperature probably varies according to the depth under the surface, the presence or absence of ground water, the nature of the rock or earth, and other conditions. In the deep tunnel which runs under Central Park, a temperature of 54 degrees was frequently observed in June and July, 1903, although the temperature outside at the same time occasionally reached 95 degrees.

In the year in which the road was opened for travel, the subway at Columbus Circle (59th Street) averaged about

53 degrees warmer than the average temperature in the streets.

In March and April the average temperatures inside and out were practically the same.

From the middle of April to the end of September, the subway at Columbus Circle was cooler than the streets.

Immediately after opening the subway to travel. From the middle of August until the opening day, October 27, 1904, the subway was gradually cooling from the highest point which it had reached in the summer. From that time to the end of the year the temperature gradually fell.

The large amount of travel which the subway experienced at the start did not visibly affect the decline of temperature which was to be expected at this season of the year.

For the first ten weeks of 1905, the subway was slightly warmer than in 1904. It is not clear that this difference was due solely to the fact that the road was in operation. The temperature out of doors was warmer than it had been the year before.

Beginning about the middle of March, or at that season when, in other years, the temperature inside and outside had been about the same, the heating due to the operation of the road first became distinctly visible. Instead of the streets becoming warmer than the subway, as had been the case the year before, the temperature inside rose also. Thenceforth a higher temperature in the subway became the rule, both winter and summer.

The summer of 1905. Throughout the investigation with which this paper is concerned, the subway was generally warmer than the streets. The only exceptions were when the outside temperature rose rapidly after a prolonged low period. This usually occurred in summer in the middle of the day, and in winter after a cold snap.

The excess of subway temperature over outside temperature increased considerably during the autumn and winter months. In the early part of July the difference between the temperature for the whole day inside and outside of the subway was less than 5 degrees. In the latter part of September it was over 10 degrees. In January, it was at some stations about 20 degrees. An average daily difference, for a week, of 30 degrees was found at one station.

Referring now to the observations of temperature made with the dry bulb thermometers of the sling psychrometers, we are prepared to obtain a better idea of the conditions which obtained in the summer after the subway was opened.

The subway grew warmer as the summer advanced. It averaged 81 degrees through July, 1905. In the week of August 4 to 10, it was 83.4 degrees. Thereafter it declined very gradually, until the latter part of September, when it was 76 degrees.

In the week of September 29 to October 5, there was a slight rise to 78 degrees, corresponding with a rise of temperature out of doors. This was followed by a more rapid decline than had occurred at any time before. Uncomfortably high temperatures were not again experienced in 1905.

During its hottest period the temperature of the subway followed the temperature of the outside air, except in the more extreme and rapid changes of the latter. This correspondence is seen to be most marked when the data for inside and outside are compared in the form of weekly and monthly averages (see Table III and Fig. 18).

The temperature in the subway for the daytime for July and August, combining the records of these two months to form an average, was 82.4 degrees; it was 76.8 degrees outside; difference, 5.6 degrees.

Highest temperatures in the summer of 1905. The highest temperature observed in the subway during the investi-

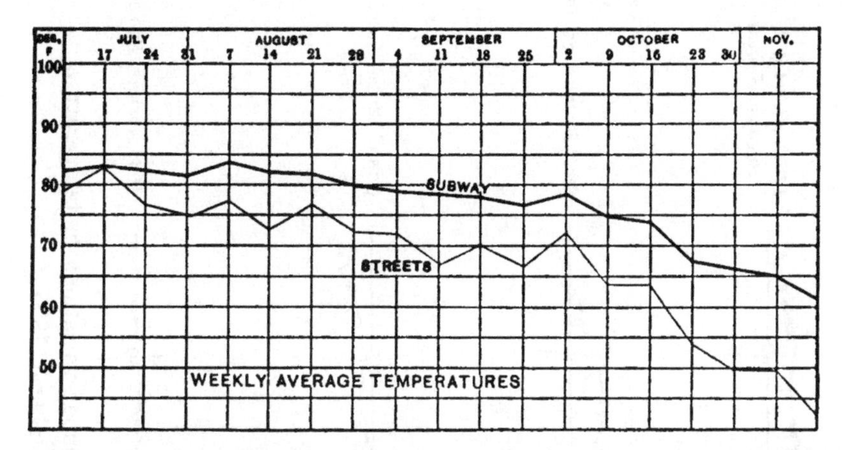

Fig. 18. Weekly average temperatures in the subway and streets from July 10 to November 13, 1905. These averages are made up of 47,476 observations.

gation was 95 degrees. This occurred at the Brooklyn Bridge station, July 18, 1905, at 3.50 P.M.

The hottest week was that of August 4 to 10, inclusive. The average daily temperature for the subway during this time was 83.4 degrees; for the outside air, 78.2 degrees; difference, 5.2 degrees.

The maximum temperature observed in the subway during this hottest week was 88.2 degrees; in the streets it was 88.2 degrees at the same time.

The warmest and coolest stations. The coolest station was Canal Street, and the warmest Astor Place. Synchronous curves of temperature for some stations and the street are given in Fig. 19.

The lowest temperature recorded at Canal Street up to January 1 was 30 degrees. The outside temperature at the same time was 14 degrees, giving a difference of 16 degrees. At the same time, the temperature at the Brooklyn Bridge and Astor Place stations was 54 degrees. Or,

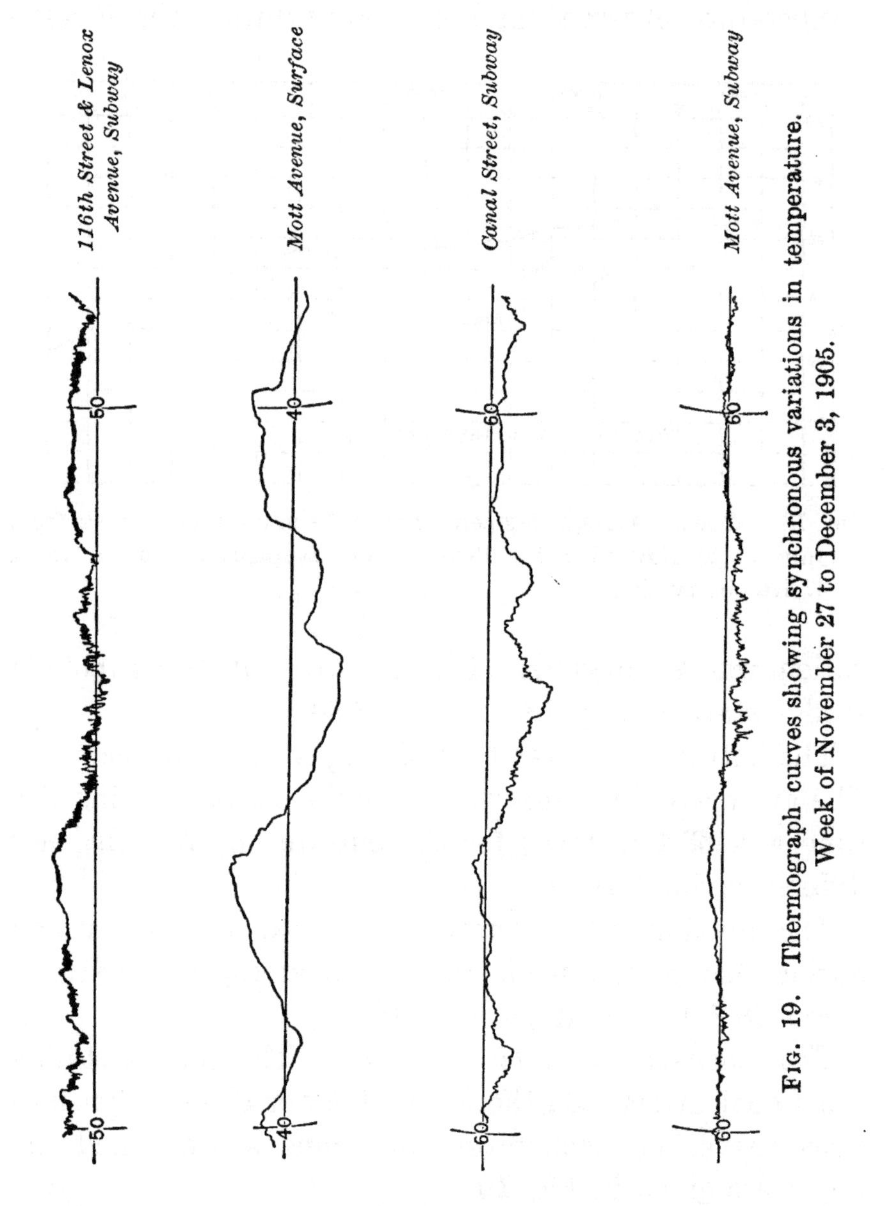

FIG. 19. Thermograph curves showing synchronous variations in temperature. Week of November 27 to December 3, 1905.

in other words, the Brooklyn Bridge station was 40 degrees warmer than the outside air, while the Canal Street station was but 16 degrees warmer.

The express stations, with the exception of 96th Street, which was exceptionally open to the outside atmosphere, were all warmer than the local stations in their vicinity.

The coolest stations were those which were most open to the street; the hottest the most closed.

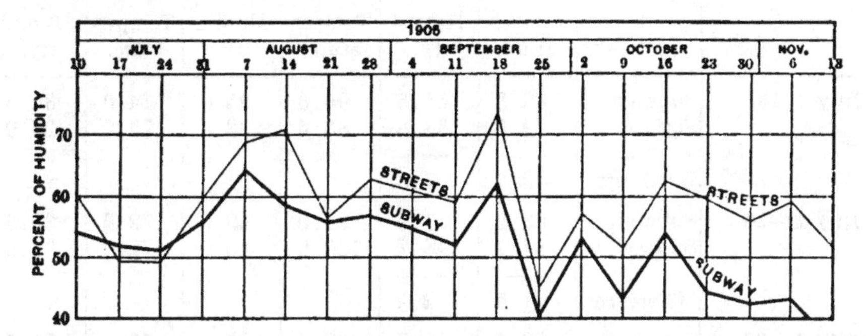

FIG. 20. Weekly average relative humidity in the subway and streets from July 10 to November 13, 1905. These averages are made up of 47,456 observations.

Humidity. The relative humidity in the subway was generally less than that out of doors, but the temperature of the dew point was higher. This is shown in Figs. 20

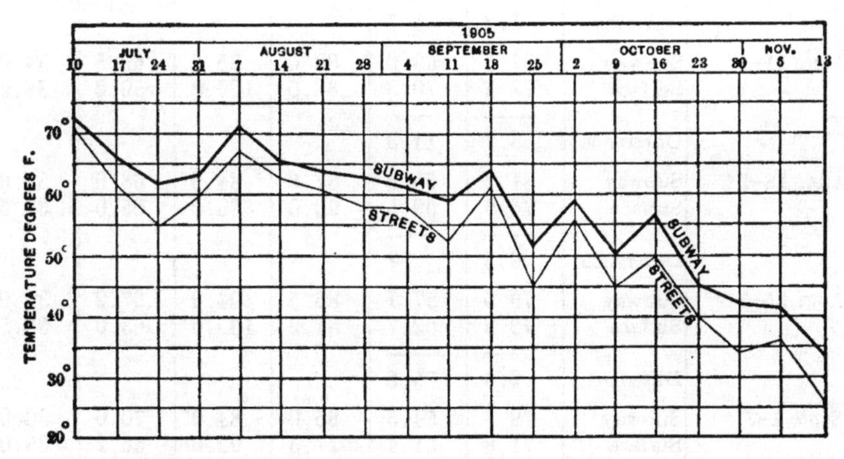

FIG. 21. Weekly average temperature of the dew point in the subway and streets from July 10 to November 13, 1905.

TABLE III

WEEKLY AVERAGE, MAXIMUM AND MINIMUM TEMPERATURE, AND RELATIVE HUMIDITY IN THE SUBWAY AND STREETS AS DETERMINED BY SLING PSYCHROMETER OBSERVATIONS FROM JULY 1 TO NOVEMBER 16, 1905. THE NUMBER OF OBSERVATIONS INCLUDED IN THIS TABLE IS 97,168

Date, 1905.	Place.	Average		Maximum		Minimum	
		Temperature.	Humidity.	Temperature.	Humidity.	Temperature.	Humidity.
July 1–13	Subway	83.3	61.5	90.0	95.0	74.0	30.0
	Surface	79.6	68.8	88.4	88.0	73.0	28.0
	Difference	3.7	7.3				
July 14–20	Subway	83.2	51.6	95.0	69.0	72.0	36.0
	Surface	84.2	48.7	100.0	80.0	71.6	32.5
	Difference	1.0	2.9				
July 21–27	Subway	82.4	50.7	90.6	85.0	70.0	31.5
	Surface	76.3	48.6	84.0	81.0	67.6	21.0
	Difference	6.1	2.1				
July 28–Aug. 2	Subway	81.2	55.7	88.6	91.0	67.0	28.5
	Surface	74.7	59.6	85.0	86.0	62.8	32.0
	Difference	6.5	3.9				
Aug. 4–10	Subway	83.4	64.4	88.2	86.0	71.5	52.5
	Surface	78.2	69.5	88.2	93.0	70.0	31.0
	Difference	5.2	5.1				
Aug. 11–17	Subway	81.5	59.0	89.0	85.0	69.5	34.0
	Surface	72 8	70.3	88.0	100.0	59.0	38.0
	Difference	8.7	11.3				
Aug. 18–24	Subway	81.2	55.8	87.8	84.5	68.0	35.0
	Surface	76.9	59.5	90.0	83.0	65.0	34.5
	Difference	4.3	.7				
Aug. 25–31	Subway	79.5	57.1	88.5	94.0	67.2	30.0
	Surface	72.0	62.7	81.9	100.0	63.0	33.5
	Difference	7.5	5.6				
Sept. 1–7	Subway	79.3	54.5	85 0	84.0	70.0	38.0
	Surface	71.8	61.4	77.5	92.0	66.2	39.0
	Difference	7.5	6.9				

TABLE III — Continued

Date, 1905.	Place.	Average		Maximum		Minimum	
		Temperature.	Humidity.	Temperature.	Humidity.	Temperature.	Humidity.
Sept. 8–14	Subway	78.4	52.0	84.5	85.0	61.0	26.0
	Surface	66.5	59.0	79.5	88.5	52.8	21.5
	Difference	11.9	7.0				
Sept. 15–21	Subway	78.1	62.2	84.2	89.0	63.2	33.5
	Surface	70.2	73.9	76.5	99.0	60.5	43.0
	Difference	7.9	11.7				
Sept. 22–28	Subway	76.0	40.5	84.2	72.	59.2	22.0
	Surface	66.0	45.0	81.2	74.5	47.8	28.0
	Difference	10.0	4.5				
Sept. 29–Oct. 5	Subway	78.0	52.6	83.8	79.0	68.8	38.0
	Surface	72.1	57.7	81.8	91.0	64.1	33.0
	Difference	5.9	5.1				
Oct. 6–12	Subway	74.4	42.6	82.2	62.0	59.0	24.0
	Surface	63.4	51.2	80.0	87.0	51.7	26.0
	Difference	11.	8.6				
Oct. 13–19	Subway	73.0	53.2	81.7	80.0	60.9	29.0
	Surface	63.0	61.6	76.8	93.0	45.8	31.5
	Difference	10.6	8.4				
Oct. 20–26	Subway	67.3	44.2	79.0	90.0	49.9	23.0
	Surface	53.7	59.5	69.1	98.5	42.2	28.0
	Difference	13.6	15.3				
Oct. 27–Nov. 2	Subway	65.6	42.3	74.1	76.	48.8	21.0
	Surface	49.4	56.0	59.0	90.5	38.9	28.5
	Difference	16.2	13.7				
Nov. 3–9	Subway	64.6	42.5	73.0	77.0	50.8	26.0
	Surface	50.1	58.9	58.5	88.0	41.5	33.0
	Difference	14.	16.4				
Nov. 10–16	Subway	60.	34.	71.0	63.0	41.8	15.0
	Surface	42.5	51.6	58.0	86.0	24.0	10.0
	Difference	18.5	16.6				

and 21. In other words, the actual weight of aqueous vapor present was greater in the subway than outside, but it appeared to be less in the subway than in the streets.

The humidity in the subway varied with the humidity out of doors.

There were no fogs nor mists in the subway. A faint haze was not uncommon.

The average relative humidity for the subway for July and August was 57.5 per cent; for the outside air, 60.6 per cent; difference, 3.1 per cent.

The greatest average relative humidity occurred during the week when the average temperature was highest. During this period the relative humidity averaged 64.4 per cent.

Condensed records of humidity are given in Table III.

CHAPTER VI

AIR OF THE NEW YORK SUBWAY, CONTINUED

CHEMICAL CONDITION OF THE AIR

THE chemical analyses of air were confined chiefly to determinations of carbon dioxide, for it was thought that no other test could give such a correct knowledge of the extent to which the air was vitiated by respiration, and none could be made on such a large scale as was wanted with so little probability of error.

Methods of analyses. — *Analyses for carbon dioxide.* The samples of air for analysis for carbon dioxide were collected in large, round-bottom flasks, varying in capacity from 2000 cubic centimeters to 2600 cubic centimeters, made especially for the purpose.

Experiments proved that titrations could be made much more accurately in round-bottom flasks than in the Erlenmeyer flasks usually employed, since the faintest pink color could readily be detected by giving the flask a rotary motion and observing the color through the depth of the liquid as it spread in a thin film upon the sides of the flask.

Each flask was provided with a well-fitting, two-hole rubber stopper, fitted with glass plugs. Before being used, the flasks were boiled with sulphuric acid to remove any free alkali which may have been present, and then carefully rinsed and standardized.

Six of these flasks were fitted into a basket, which was carried by the collector. A large football pump, with the valve reversed, to pump air from the flasks, was employed; and this, with about 10 feet of rubber tubing, a ther-

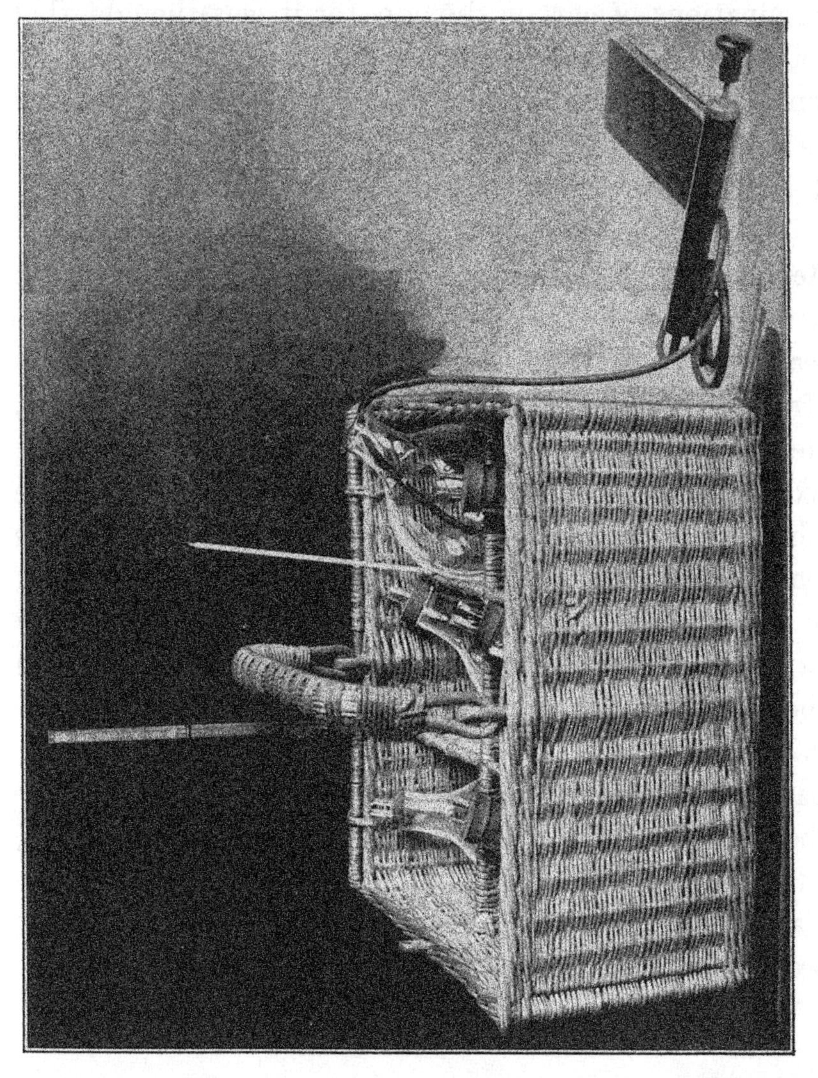

Fig. 22. Apparatus for carbon dioxide analyses, flasks, air pump and other apparatus used in collecting samples of air.

mometer, and a notebook, completed the collector's outfit. This apparatus is shown in Fig. 22.

When it was desired to collect a sample, the basket was opened and the stopper removed from one of the flasks. The free end of the rubber tubing was then inserted into the flask as far as the bottom. This done, the operator removed to a distance with his pump and pumped air from the flask for about four minutes. This amount of pumping was capable of removing about eight times the volume of

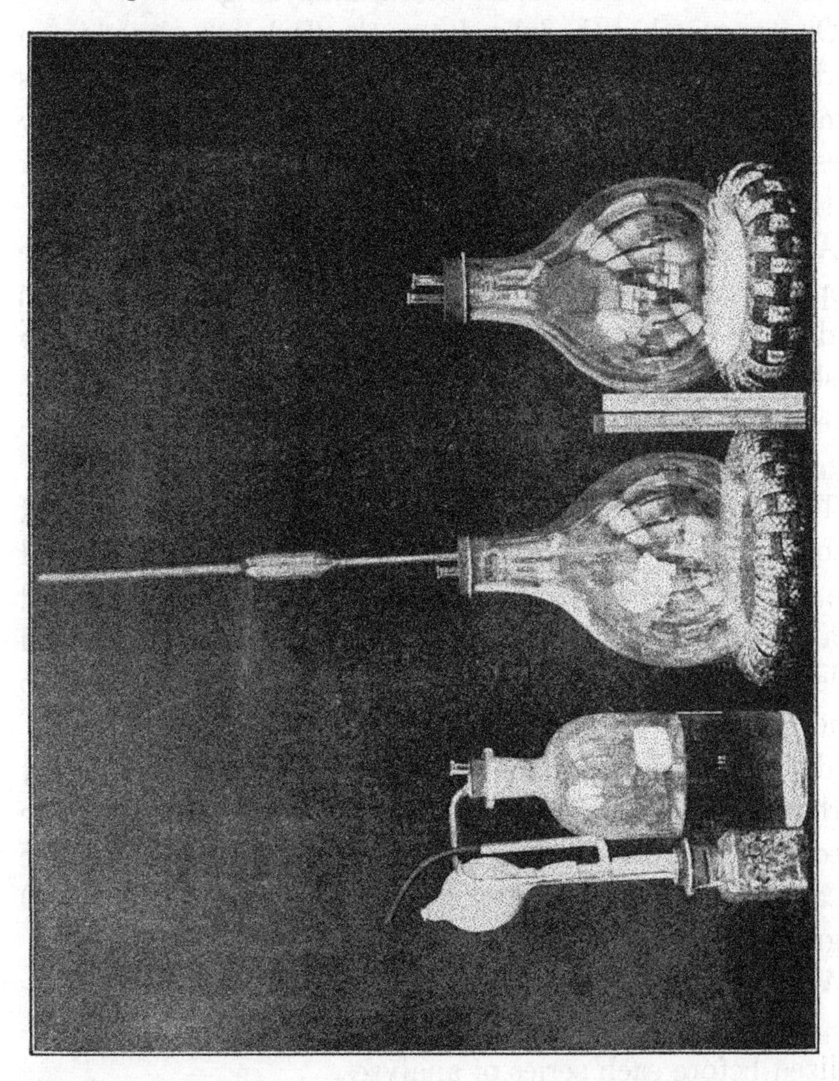

Fig. 23. Apparatus for carbon dioxide analyses. Method of adding barium hydroxide to the flasks of air.

air in the flask, and provided for the collection of a proper sample.

After the sample was collected, the stopper was replaced in the flask, the temperature noted, and the data observed which were to fix the identity of the sample.

The samples were usually analyzed on the day of collection or the day after. Experiments showed that no appreciable change took place by allowing a sample to stand for twenty-four hours before analyzing.

On its arrival at the laboratory, the flask was placed in an upright position on a suitable rest and one of the glass rods removed from the stopper. The stopper was then gently pressed down until it reached a point marked on the neck of the flask at which its capacity had been calibrated.

A pipette was next inserted through the hole in the stopper and 20 cubic centimeters of standard barium hydroxide, to absorb the CO_2, allowed to flow into the flask with a few drops of plenolphthalein. (See Fig. 23.) The solution was allowed to stand, with occasional shaking, for one hour.

The delivery tube of a special burette was then inserted into the flask through one of the holes in the stopper, and the excess of barium hydroxide titrated with standard oxalic acid. (See Fig. 24.) The glass rod in the second hole of the stopper was removed from time to time to relieve the pressure. One cubic centimeter of this oxalic acid was equivalent to one-tenth of a cubic centimeter of carbon dioxide. From the quantity of barium hydroxide used, the amount of carbon dioxide in the original sample of air was calculated, the volume being reduced to 0 degrees Centigrade and 760 millimeters pressure.

Although the barium hydroxide solution remained practically constant from day to day, it was always standardized before each series of analyses.

The oxalic acid employed for making up the standard solution was tested by titrating with standard potassium permanganate and found to be satisfactorily pure.

The bottle in which the standard solution of barium

hydroxide was kept was provided with a safety bottle containing pumice and caustic soda and a rubber bulb, which forced the liquid up into the pipette.

FIG. 24. Apparatus for carbon dioxide analyses. Method of titration with oxalic acid after the absorption of the carbon dioxide by barium hydroxide.

The burette used for titration was especially made for the purpose, and had a delivery tube 9 centimeters long, which in use projected well into the flask of air.

Inasmuch as the manner in which the sample was collected gave room for some error if the collector was inattentive or careless in his work, care was taken to collect check samples from time to time by the help of other assistants. This error was further guarded against by entrusting the collections only to persons who, by age and training, seemed certain to use proper care. One of the collectors held the rank of assistant engineer under the Municipal Civil Service Commission of the city. Another was an analytical chemist of long experience.

A sample of air which was taken to check the work of a collector gave the following results:

Conditions of experiment.	Parts CO_2 per 10,000 volumes of air.
Collector's sample	3.54
Samples collected as a check	3.48
Difference	.06

Experiments were made to determine whether or not long standing of the barium hydroxide solution in contact with the flask would have any effect on the results. For this purpose four samples of outside air were collected at the same time and place, and the barium hydroxide solution allowed to remain in the flask for periods of from one to four hours. The results follow:

Conditions of experiment.	Parts CO_2 per 10,000 volumes of air.
Titrated at the end of one hour	3.48
Titrated at the end of two hours	3.47
Titrated at the end of three hours	3.49
Titrated at the end of four hours	3.48

Experiments to ascertain whether or not any difference in the results would be obtained by adding the barium hydroxide to the flask at the time of collection gave results which follow:

Conditions of experiment.	Parts CO_2 per 10,000 volumes of air.
Solution added at time of collection	4.33
Added four hours after collection	4.36

Another set of samples was taken and analyzed as follows:

No.	Conditions of experiment.	Parts CO_2 per 10,000 volume. of air.
1	Added barium hydroxide solution at time of collection	2.09
2	Added barium hydroxide solution at time of collection	3.02
3	Added barium hydroxide solution after eighteen hours	3.02
4	Added barium hydroxide solution after eighteen hours	3.11
5	Added barium hydroxide solution after eighteen hours (paraffined stopper)	3.13
6	Added barium hydroxide solution after eighteen hours (paraffined stopper)	3.22

Another set of samples was analyzed as follows:

No.	Conditions of experiment.	Parts CO_2 per 10,000 volumes of air.
1	Added barium hydroxide at time of collection . . .	3.40
2	Added barium hydroxide at time of collection . . .	3.56
3	Added barium hydroxide after twenty-four hours. .	3.44
4	Added barium hydroxide after twenty-four hours. .	3.57

Analyses for oxygen. About eighty samples of air were analyzed for oxygen. The difference between the amount present in the subway and in the streets seemed so slight and uninstructive that the determinations were soon discontinued as a routine procedure.

The samples of air were collected in glass tubes of 300 cubic centimeters capacity, closed at each end by glass stopcocks. The tubes were filled with oxygen-free water at the laboratory and taken to the point where the samples were to be collected. There the cocks were opened and the water allowed to flow out, the desired sample of air taking its place. The cocks were then closed and the sample taken to the laboratory for analysis.

The oxygen was determined by absorption with phosphorus, according to the method of Lindemann. Hempel

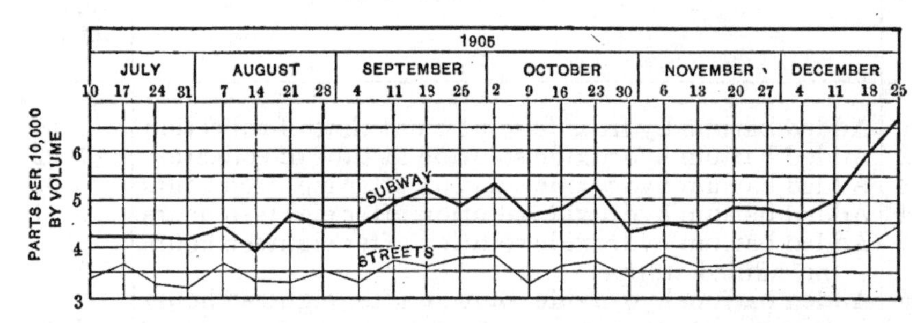

FIG. 25. Weekly average carbon dioxide in the subway and streets from July 10 to December 25, 1905. The number of determinations included in this figure was 1,772.

burettes and a Hempel pipette, constructed for solid absorbents, were employed. The phosphorus was specially prepared, and the pipette was kept covered to protect it from the action of light. The pipette, when in use, was immersed in water to maintain a constant temperature, and thus obviate any inaccuracy which might have been caused by temperature changes. (See Fig. 26.)

Carbon dioxide results. The carbon dioxide analyses produced results from which the author derived the following conclusions:

FIG. 26. Apparatus for determining the proportion of oxygen in air. The sample of air to be analyzed was collected in the receptacle which lies on the floor.

The average amount of carbon dioxide in the subway was a little larger than the amount in the air of the streets.

The average of all results was, for the subway, 4.81 volumes per 10,000 volumes of air, and for the streets, 3.67; difference, 1.14. This difference must be regarded as very slight. (See Fig. 25.)

The frequency with which the air was renewed could not be accurately calculated, for the reason that the number of passengers traveling in the subway was not known.

At no time or place was the amount of carbon dioxide large.

The greatest amount of carbon dioxide found in the subway was 8.89. This occurred in one of the tunnels between the Grand Central station and the 33d Street station, on December 27, 1905, at 6.02 P.M. At the same time there was a block in the traffic, during which trains were stalled at all points in the vicinity. At the adjoining stations of 33d Street and Grand Central, the carbon dioxide was higher than usual at the same time, the amount at 33d Street being 7.84 and at Grand Central 7.87.

The carbon dioxide in the subway varied according to season, hour, place where the sample was collected, and other circumstances.

Season. There was more carbon dioxide found in the autumn than in the summer or winter. (See Fig. 27.) It seemed likely that this was explainable largely on the ground that many more passengers were carried in autumn than in summer, and that in winter there was more wind in the streets and the subway, increasing the amount of ventilation.

Hourly variations. The amount of carbon dioxide varied in the subway at different hours of the day. (See Fig. 28.) These irregularities corresponded with the irregularities in the amount of travel which took place at different hours.

It is interesting to note that periodic changes in the

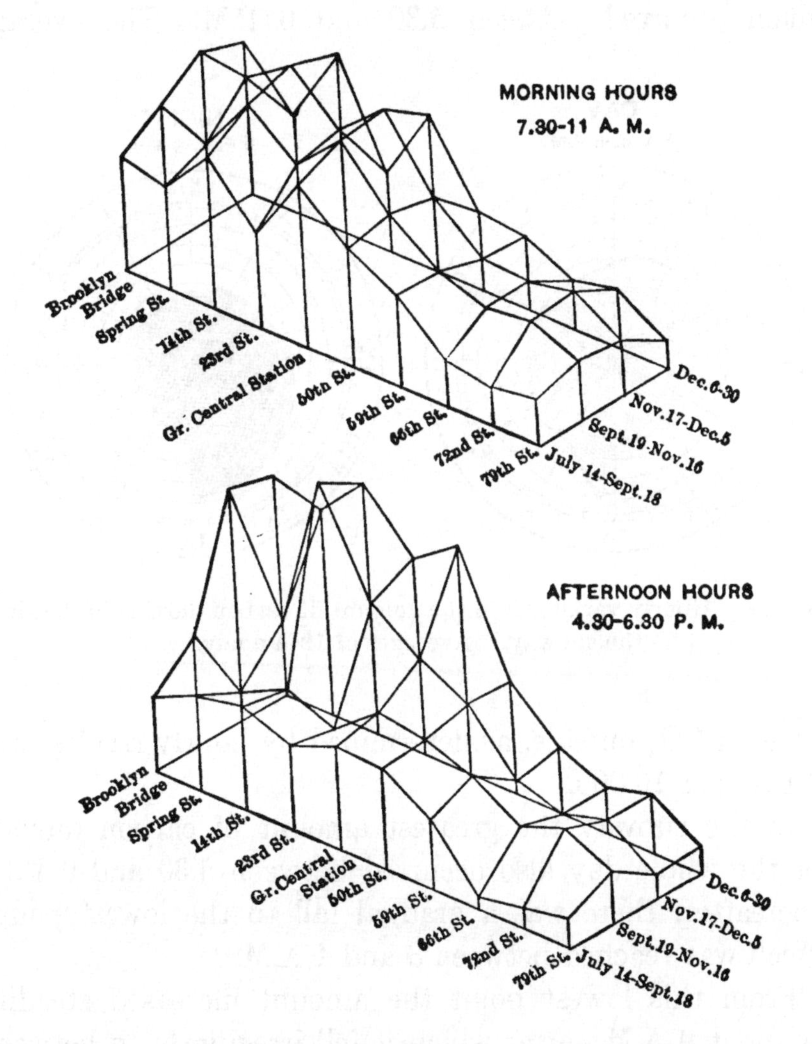

FIG. 27. Carbon dioxide at different stations at different seasons during the hours of maximum travel in the morning and afternoon.

amount of carbon dioxide occurred in the streets. In the streets the carbon dioxide was highest between 5.30 and 6 P.M., and lowest between 1 and 3 A.M. The amount increased from a minimum in the early morning hours to about 9 A.M. After this there was a fall to about 1.30 P.M., followed by a rise to the highest point of the day, which occurred between 5.30 and 6 P.M. The average

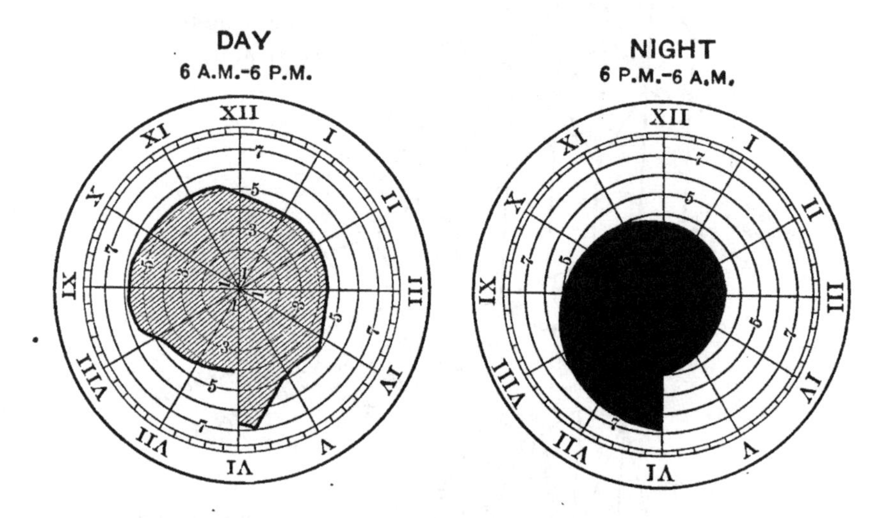

FIG. 28. Hourly variations in the amount of carbon dioxide in the air of the subway. Averages of 1244 analyses.

range of CO_2 outside, as determined by hourly results, was .8 part per 10,000.

In the subway the greatest amount of carbon dioxide for the whole day also occurred between 5.30 and 6 P.M. Thereafter, there was a gradual fall to the lowest point, which was reached between 3 and 4 A.M.

From this lowest point the amount increased steadily to about 9 A.M., after which it fell irregularly to between 1 and 2 P.M.

The average for the whole day agreed closely with the average between 1 and 3 P.M.

In the late afternoon there was a rapid rise to the maximum for the day, which was reached at about 5.30 P.M.

TABLE IV. — SIMULTANEOUS DETERMINATIONS OF CARBON DIOXIDE AT AND BETWEEN STATIONS IN THE SUBWAY. THE NUMBER OF ANALYSES INCLUDED IN THIS TABLE IS 442.

Date, 1905.	Point of observation.	Carbon Dioxide.	
		Found.	Excess.
July 31–Aug. 4	Between Fulton Street and Brooklyn Bridge stations	4.26	
	Average for Fulton Street and Brooklyn Bridge stations	4.07	.19
	Between Brooklyn Bridge and Worth Street stations	4.35	
	Average for Brooklyn Bridge and Worth Street stations	4.37	−.02
	Between Worth Street and Canal Street stations	4.13	
	Average for Worth Street and Canal Street stations	4.29	−.16
	Between Canal Street and Spring Street stations	4.18	
	Average for Canal Street and Spring Street stations	4.47	−.29
	Between Spring Street and Bleecker Street stations	4.28	
	Average for Spring Street and Bleecker Street stations	4.40	−.12
	Between Bleecker Street and Astor Place stations	4.57	
	Average for Bleecker Street and Astor Place stations	4.61	−.04
July 29–Aug. 5	Between Astor Place and 14th Street stations	4.68	
	Average for Astor Place and 14th Street stations.	4.77	−.09
	Between 14th Street and 18th Street stations	4.46	
	Average for 14th Street and 18th Street stations	4.57	−.11
	Between 18th Street and 23d Street stations	4.17	
	Average for 18th Street and 23d Street stations	4.27	−.10
July 28–Aug. 2	Between 28th Street and 33d Street stations	4.12	
	Average for 28th Street and 33d Street stations	3.99	.13
	Between 33d Street and Grand Central stations	4.48	
	Average for 33d Street and Grand Central stations	4.23	.25
July 28–Aug. 1	Between Grand Central and Times Square stations	4.73	
	Average for Grand Central and Times Square stations	4.46	.27
	Between Times Square and 50th Street stations	4.44	
	Average for Times Square and 50th Street stations	4.27	.17
	Between 50th Street and 59th Street stations	4.07	
	Average for 50th Street and 59th Street stations	3.99	.08
Dec. 14–23	Between Fulton Street and Brooklyn Bridge stations	6.31	
	Average for Fulton Street and Brooklyn Bridge stations	6.21	.10
Dec. 14–26	Between Canal Street and Spring Street stations	6.06	
	Average for Canal Street and Spring Street stations	5.92	.14
Dec. 15–22	Between 14th Street and 18th Street stations	6.74	
	Average for 14th Street and 18th Street stations	6.55	.19
Dec. 26–27	Between 33d Street and Grand Central stations	7.21	
	Average for 33d Street and Grand Central stations	6.82	.39
Dec. 15–22	Between Grand Central and Times Square stations	6.55	
	Average for Grand Central and Times Square stations	6.21	.34
Dec. 13–21	Between 66th and 72d Street stations	4.78	
	Average for 66th Street and 72d Street stations	4.75	.03
Dec. 28	Under Central Park, south of 110th Street station	6.18	
	At 110th Street station	5.45	.73
Dec. 29	Under Harlem River	6.30	
	Mott Avenue station	4.58	1.72
Mean CO_2 between stations		5.05	
Mean CO_2 at stations		4.87	.18

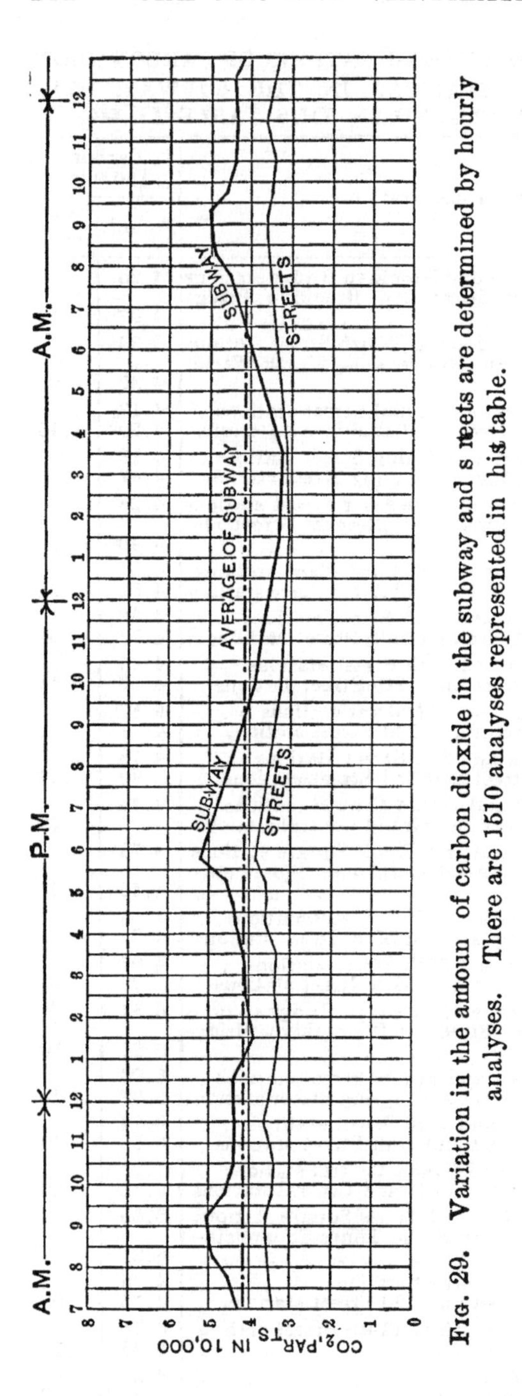

FIG. 29. Variation in the amount of carbon dioxide in the subway and streets are determined by hourly analyses. There are 1510 analyses represented in this table.

The difference between the least and greatest amounts of carbon dioxide was, according to these hourly averages, two parts per 10,000.

Figure 29 shows the variation in the amount of CO_2 in the subway and streets as determined by hourly results.

Differences in different parts of the subway. There was more carbon dioxide at express stations than at local stations, except at the especially open express station at 96th Street.

Among the principal express stations, the largest average amount of carbon dioxide was found at 14th Street. Then came the Brooklyn Bridge, Grand Central, 72d Street, and 96th Street stations, in the order named.

There was more carbon dioxide between stations than at the adjoining stations, although in most cases this difference was very little. The results of 442 analyses showed this average difference to have been .18 part per 10,000, with a range from .29 to 1.72. Data relating to this subject are given in Table IV.

Marked differences occurred in the amount of carbon dioxide found at points above and below 50th Street. The average of all results which were capable of being taken into account to show this difference demonstrated that the air from 50th Street uptown was much purer than the air from 50th Street downtown. This is shown in Table V.

TABLE V.

CARBON DIOXIDE IN THE SUBWAY NORTH AND SOUTH OF FIFTIETH STREET. THE NUMBER OF ANALYSES INCLUDED IN THIS TABLE IS 1,182.

FROM FIFTIETH STREET DOWNTOWN.

1905.	Brooklyn Bridge.	Spring Street.	Fourteenth Street.	Twenty-third Street.	Grand Central.
July 14–Sept. 18	1.14	1.08	1.39	0.97	1.46
Sept. 19–Nov. 16	1.34	1.03	1.50	1.17	1.61
Nov. 17–Dec. 5	2.17	1.51	2.58	1.77	1.91
Dec. 6–Dec. 30	2.01	1.62	2.37	1.38	2.08
	1.66	1.31	1.96	1.32	1.76

FROM FIFTIETH STREET UPTOWN.

	Fiftieth Street.	Fifty-ninth Street.	Sixty-sixth Street.	Seventy-second Street.	Seventy-ninth Street.
July 14–Sept. 18	1.20	.88	.31	.32	.42
Sept. 19–Nov. 16	1.26	.97	.71	.86	.61
Nov. 17–Dec. 5	2.27	.71	.63	.82	.42
Dec. 6–Dec. 30	1.02	.75	.52	.65	.30
	1.44	.83	.54	.66	.44

Differences in the elevation above the pavement at which samples were taken made little difference in the amount of CO_2 found in the subway. These differences are indicated in Table VI.

TABLE VI

CARBON DIOXIDE AT DIFFERENT ELEVATIONS ABOVE THE PLATFORMS OF SUBWAY STATIONS

Date, 1905.	Place.	Time.	Two feet.	Four feet.	Six feet.	Eight feet.	Ten feet.	Average.
Aug. 17	14th Street station . .	2.00– 2.15p	4.01	3.48	4.11	4.24	...	4.05
Aug. 18	Grand Central station	9.45–10.10a	5.35	4.83	4.73	4.71	4.73	4.87
	Brooklyn Bridge station	1.40– 2.15p	4.21	4.41	4.36	4.18	4.18	4.27
Aug. 19	Grand Central station	9.35–10.00a	4.84	4.61	4.59	4.69	4.71	4.69
Aug. 22	14th Street station . .	10.45–11.10a	4.72	4.97	4.69	4.90	4.60	4.78
Aug. 23	Grand Central station	10.00–10.30a	4.63	4.67	4.74	4.59	4.65	4.66
Aug. 24	Brooklyn Bridge station	9.55–10.25a	4.91	4.92	4.64	4.50	4.74	4.74
Mean CO_2 at different heights			4 67	4.61	4.55	4.54	4.60	4.58
Mean of all observations			4.58	4.58	4.58	4.58	4.58	4.58
Departure of mean for each height from mean of all observations			+.09	+.03	−.03	−.04	+.02	...

The CO_2 in the air of the cars in summer, when the windows and front doors were open and the travel comparatively light, was not far different from the air of the subway itself.

Oxygen results. The samples of air which were analyzed for oxygen were collected from 9.30 A.M. to 5.30 P.M., between the Brooklyn Bridge and 96th Street stations.

The average amount of oxygen found in the air of the streets was 20.71 per cent; in the subway, 20.60 per cent; difference, .11 per cent. The least amount found in the subway was 20.25 per cent.

BACTERIAL CONDITION OF THE AIR

The studies concerning the microörganisms in the subway related chiefly to the number and origin of the bacteria and molds. It was not practicable within the time and scope of the investigation to determine the various species of bacteria present, but the principal sources of many of them were investigated indirectly with fairly satisfactory results.

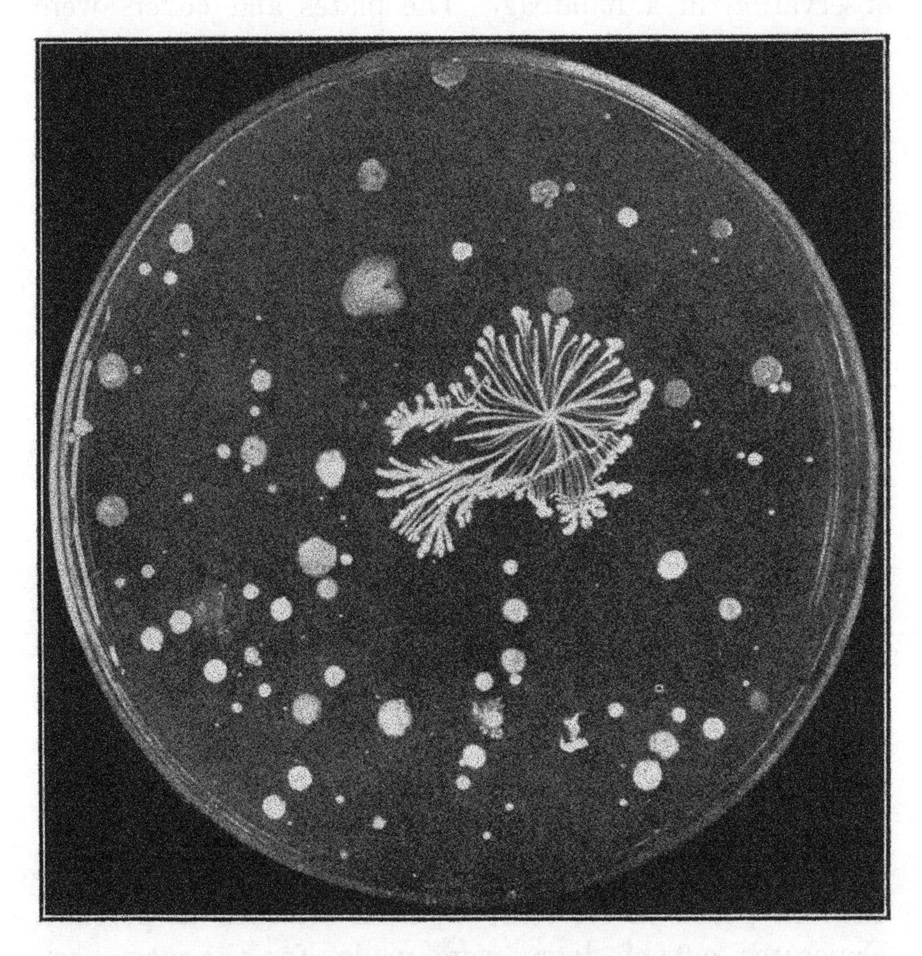

Fig. 30. Colonies of bacteria grown on a plate exposed to the air for fifteen minutes at the Grand Central Subway Station.

Quantitative methods of analysis. The bacteria were collected by allowing them to settle from the air on circular plates, or Petri dishes, $3\frac{1}{2}$ inches, or about 9 centimeters, in diameter, containing a standard agar culture medium, and by collecting them from the air by means of sand filters.

The plate method. The plates containing the culture medium were carried from the laboratory to the points of observation in a handbag. The plates and covers were fastened together by means of elastic bands, each pair separated from other pairs by sterilized towels.

To make an observation, the plates were taken from the handbag and placed, usually, upon a bench about 18 inches from the pavement. The covers were then removed and kept off for fifteen minutes, or a fraction of this time, depending upon the observer's judgment of the numbers which would probably be found.

During the exposure of the plates, the observer removed to a distance and made the notes necessary to identify the observation.

The covers were then replaced, secured to their respective dishes by the rubber bands, and returned to the laboratory in the handbag. The plates were put into the incubator within two hours after exposure.

The plates were incubated at a temperature of 37 degrees Centigrade for forty-eight hours. The colonies of bacteria and molds which developed in this time were then counted. Care was taken to separate the two.

Most of the exposures were made at the subway stations at points as far removed from draughts as possible. Exposures out of doors were made, for the most part, on the line of the subway beyond the influence of subway air, and at an elevation of about 3 feet above the sidewalk.

In all, about 2800 exposures were made.

The agar culture medium was prepared with Liebig's extract of beef and 5 per cent agar.

The reaction of the medium, after preliminary trials to ascertain the optimum, was fixed at ½ to 1 per cent acid to phenolphthalein. In preparing the plates for use, 10 cubic centimeters of the agar medium was poured, and the plates then put into the incubator for twenty-four hours. Plates which developed colonies in this time were contaminated and not used.

All the bacteriological plates were exposed in duplicate. The results reported were averages of the counts of two plates in every instance.

Preliminary trials of different media and at different periods of incubation showed that a gelatin medium kept for several days at or below what is generally termed "room temperature" would often develop more colonies than an agar medium developed at body temperature. There were probably several reasons for this: The higher temperature of the body undoubtedly kept many delicate air organisms from growing. Gelatin was a more favorable solidifying agent than agar. An incubation period of more than forty-eight hours seemed to be essential to the growth, to visible colonies, of some bacteria in artificial culture media.

The chief objection to the gelatin-room-temperature method was that gelatin melted at a temperature which made it unsuitable for summer use, became too easily liquefied by certain bacteria, and was likely to be too rapidly overgrown with molds.

The procedure adopted seemed to be the most practical under the circumstances. It had also the merit of being capable of duplication at any time and place with a considerable degree of accuracy.

The numbers of bacteria which actually existed in the air must have been largely in excess of the numbers found, but it was not feasible to ascertain how great was this difference.

Colonies of bacteria grown in a plate exposed for fifteen minutes to the air at the Grand Central subway station are shown in Fig. 30.

The filter method. The collection of microörganisms by the use of filters provided means for estimating the numbers of bacteria and molds recoverable from a measured volume of air. After a considerable amount of preliminary work, the method adopted was substantially that employed by Sedgwick, Prudden, and others.[1]

The filters consisted of glass tubes about 13 centimeters long and 0.5 centimeters inside diameter. At one end the filtering material was held in place by a small plug of wire gauze, and at the other, when the filter was not in use, by a plug of cotton.

Various filtering materials were tried in experiments preliminary to the adoption of a standard method, especially the sugar medium of Sedgwick, proposed before the Society of Arts in 1888.[2] In the subway work the advantages of a soluble medium were more than offset by the care with which the sugar required to be sterilized and kept dry.

The filtering medium finally adopted was sand. The depth of sand was about 5 centimeters. Most of the grains were about half a millimeter in diameter. The particles were largely quartz. Two filters were always arranged in tandem. (See Fig. 32.)

Air was made to pass through the filters, in most cases by means of an exhaust pump of accurate construction, whose action had been carefully tested. Sixty-six strokes pumped

[1] Firth, "Studies from the Department of Pathology, College of Physicians and Surgeons," New York, Volume VII, 1900. (See Fig. 32.)

[2] Sedgwick and Tucker, "Methods for the Biological Examination of Air," *Proceedings of the Society of Arts*, Boston, 1888.

20 liters, and this was the amount usually pumped in each case.

Where it was not feasible to use the air pump, a brass vacuum cylinder of about 600 cubic inches, or 10 liters,

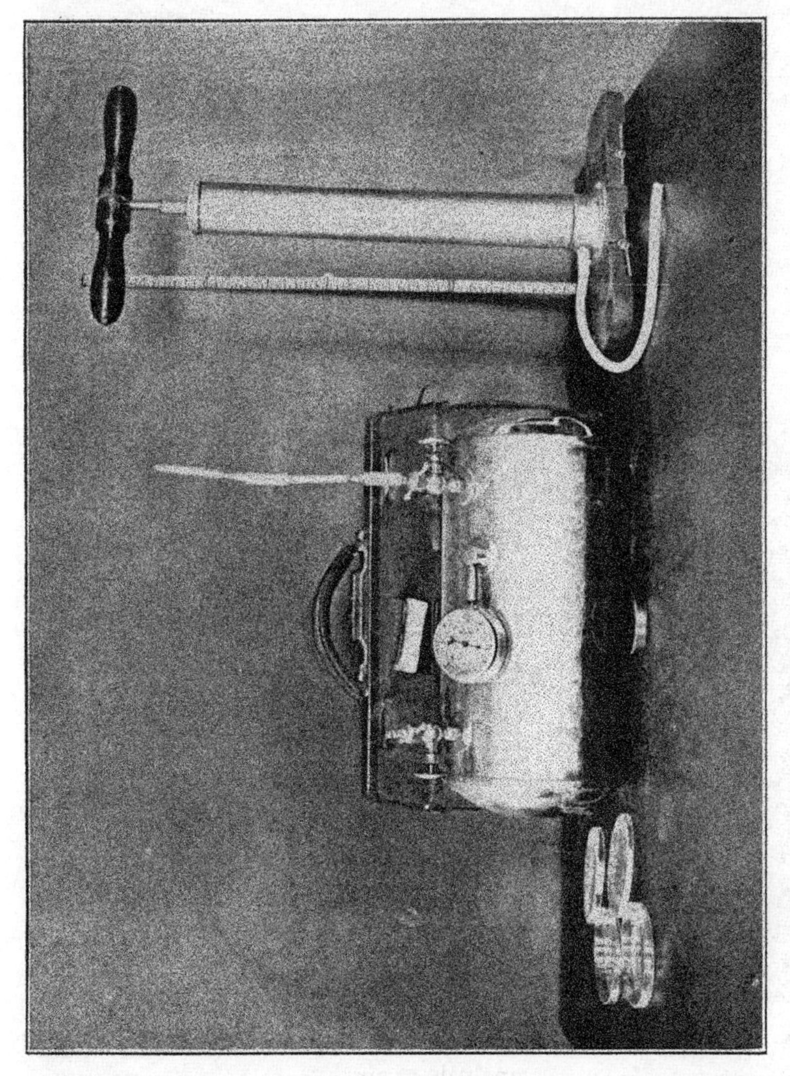

FIG. 31. Vacuum cylinder with sand filter attached, air pump and other apparatus used in collecting bacteria from air in crowded places. In use the cylinder fits inside the handbag.

capacity, fitted with a pressure gauge and suitable stopcocks, was employed, as devised by Prudden. The cylinder was 266 centimeters long and 177 centimeters in diameter, and was fitted into a neat leather bag with suitable openings, through which the gauge could be read and the filters connected. (See Fig. 31.)

Before using the cylinder, the air was exhausted from it

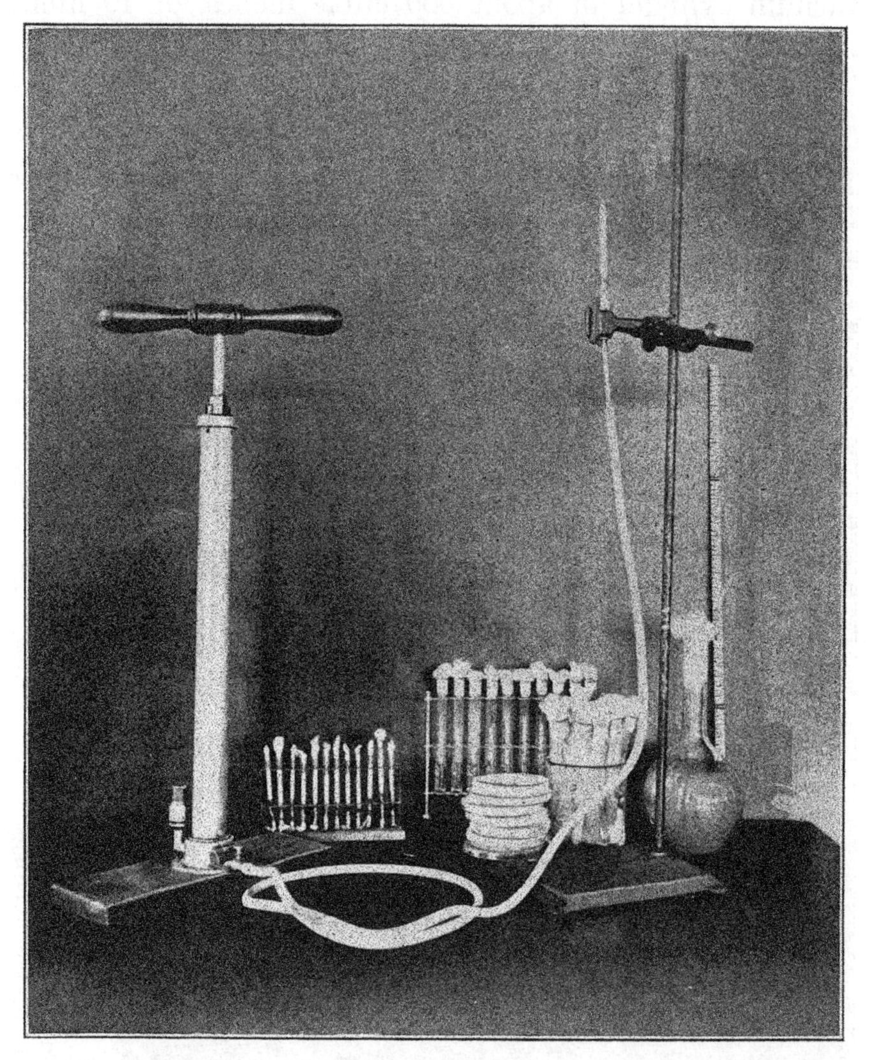

FIG. 32. Air pump, sand filters and other apparatus for bacteriological analysis of air.

and the stopcocks closed. It was then placed in the hand-bag and taken to the place where the observation was to be made. The filters were then connected to the cylinder by means of short rubber tubing, the pressure gauge read, and the air allowed to flow into the cylinder through the filters. When the desired quantity of air had been filtered, the cocks were closed, the filters removed, and the gauge read.

Upon returning to the laboratory the sand was emptied from the first filter into a test tube containing 10 cubic centimeters of sterile water, and thoroughly agitated.

The water, with the organisms which had been rinsed from the sand, was then plated in agar and incubated for forty-eight hours at 37 degrees Centigrade and the colonies counted.

The second filter was treated in the same way. Molds were excluded from the count.

Another method, and the one which was adopted finally, discarded the use of the distilled water. The sand from a filter was poured into a tube of melted agar, the tube agitated and then poured into a Petri dish, leaving the sand behind. A second tube of agar was then poured upon the sand remaining in the first tube. This was then agitated and the agar and sand both poured into a second Petri dish. The two Petri dishes were then incubated.

Most of the bacteria found were caught in the first filter.

Additional bacterial work. Beside the routine estimates of the number of bacteria recovered from the air, special studies were made of the length of life of the pneumococcus in the subway, the numbers of bacteria in subway and other dusts (see Fig. 33), the action of lubricating oil upon bacteria, the kinds of molds present, and the efficiency of various commercial deodorants and germicides intended for subway use.

The longevity of the pneumococcus. The longevity of the pneumococcus was tested by drying upon a flat piece of broken trap rock from the roadbed about 20 cubic centimeters of sputum from a patient suffering from lobar pneumonia in the congestive stage. The presence of the

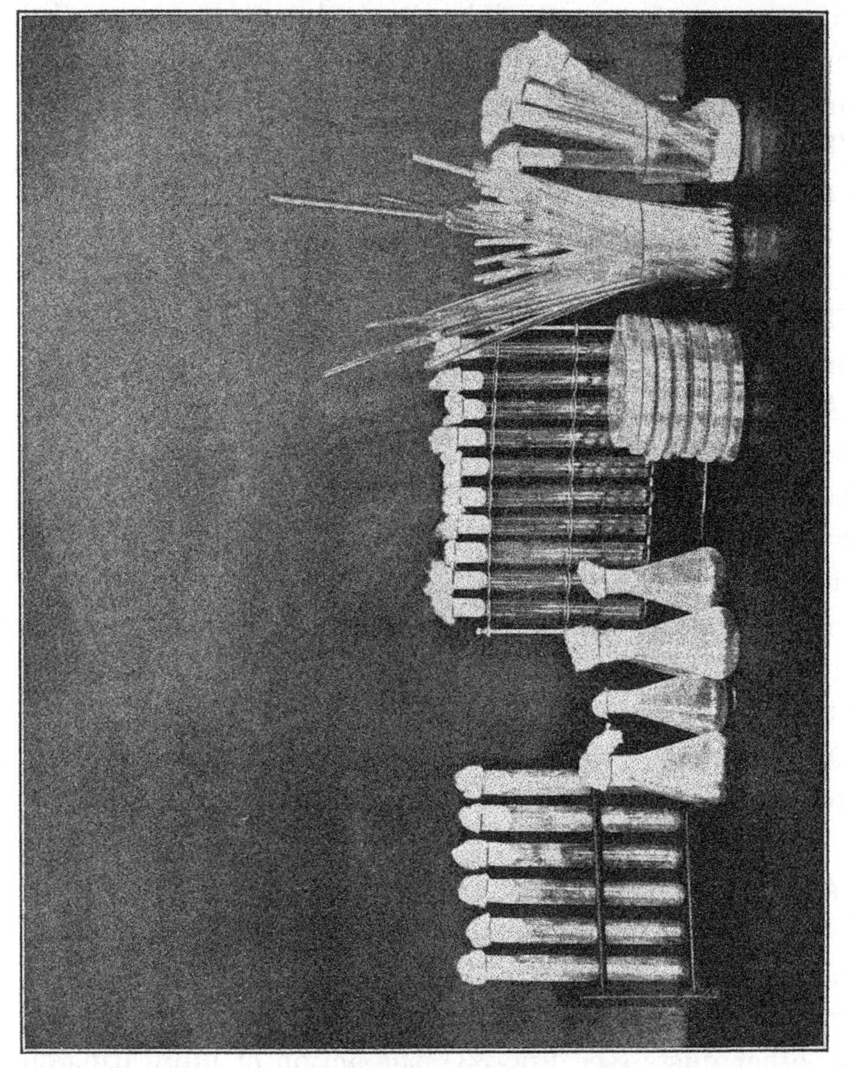

FIG. 33. Apparatus for the bacteriological analysis of accumulated dust.

pneumococcus in a viable condition was tested in the beginning of the experiment and at the end of every two or three days. The vitality was determined by culture methods and by inoculating small portions of the moistened sputum into guinea pigs.

The rock with the dried sputum was taken into the subway and set upon the top of a clean iron beam at a little distance from the end of a station platform. Small parts of the mass were removed from day to day for examination, but the main body of sputum was not taken out of the subway until the end of the experiment.

The numbers of bacteria in dust. Specimens of dust which had gathered upon recently painted beams and other clean, dry surfaces were collected in large, sterile test tubes and Erlenmeyer flasks and taken to the laboratory. Here a small, weighed portion of the dust was mixed with from 50 to 500 cubic centimeters sterile water. One cubic centimeter, and fractions thereof, of the water were then sown in the agar culture medium already described. The agar was incubated for forty-eight hours at 37 degrees Centigrade and the colonies counted. From these counts the numbers of bacteria per gram of dust were calculated.

Action of oil on bacteria. The possibility that the lubricating oil which was copiously used in the first few months after opening the subway might have a germicidal or cultural effect upon bacteria was inquired into.

Specimens of oily wood, stone, and refuse paper from the tracks were taken to the laboratory and examined for the numbers of bacteria which they contained. The oil itself was tested for its action upon various species of bacteria.

Kinds of molds. Although it was not feasible to study the kinds of bacteria, because of the great similarity in their forms and cultural behavior, it was practicable to determine a few of the molds present. This was done by fishing from the mixed colonies of molds and bacteria which developed upon the plates exposed in the subway.

From specimens obtained in this way, pure cultures were made upon various media suitable for the propagation of the molds.

No attempt was made to make this work exhaustive, since it seemed doubtful whether a complete knowledge of the numbers and kinds of molds in the subway, had it been obtainable, would have thrown much light upon the problems in hand.

Efficiency of deodorants and disinfectants. The efficiency of different chemical compounds advocated by various persons to purify the air of the subway was examined into. They were tested both with reference to their capacity for destroying unpleasant odors and bacteria.

The capacity of these compounds to destroy odors was tested in the laboratory by producing unpleasant odors in suitable closed chambers, and then introducing the compounds in different amounts and under different circumstances of temperature and humidity.

The disinfectant properties of the compounds were tested by allowing them to act upon bacteria in colonies on culture media, in films and emulsions, on toothpicks, and on freshly impregnated cotton threads. The germs used in these experiments were the typhoid bacillus and the colon bacillus, the pneumococcus and the Staphylococcus pyogenes albus. The laboratory procedures were such as are usually employed in careful work of this kind.

Results. A careful examination of the bacterial data collected in these studies, excepting the data which relate to the dust, led the author to the following conclusions:

Numbers of bacteria in subway and streets compared. There were, on an average, more than twice as many bacteria found in the air of the streets as in the air of the subway, excepting after rains, when fewer were found outside than inside.

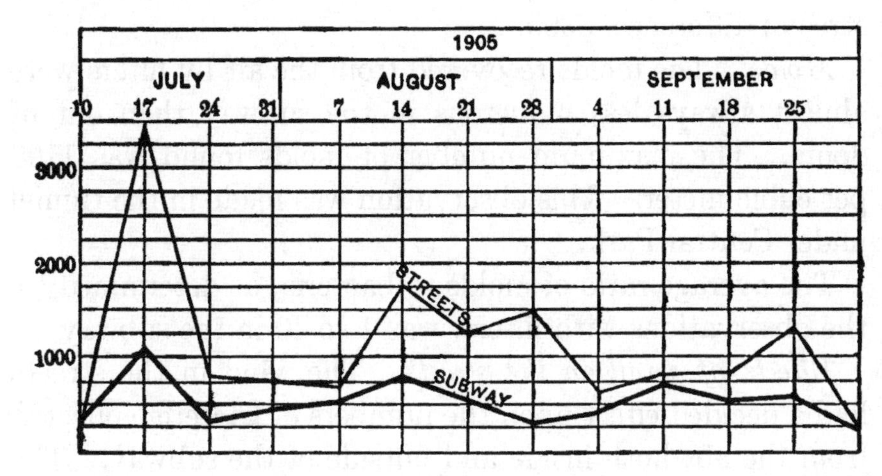

Fig. 34. Average numbers of bacteria which subsided from the air per square foot per minute, as determined by the plate method, in the subway and streets from July 10 to October 2, 1905. The number of samples represented is 2742.

The average numbers of bacteria which settled from the air in fifteen minutes, and were subsequently enumerated,

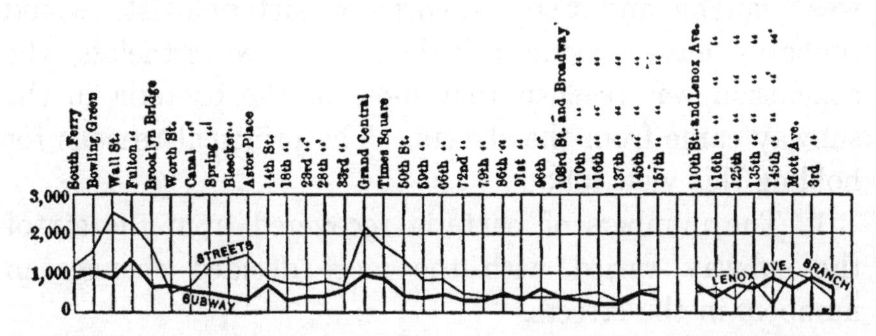

Fig. 35. Average numbers of bacteria which settled from the air upon each square foot per minute, at different subway stations and in the streets, and were subsequently counted. The number of samples represented is 2753.

were, in the subway, 500; outside, 1157; difference, 657. (See Figs. 34 and 35.)

The average number of bacteria found by filtering the air was 3200 per cubic meter in the subway and 6500 in the streets; difference, 3300.

Molds. The molds recovered from the air by filters were almost always less numerous in the subway than out of doors. The maximum number of molds found was 1100 per cubic meter. This observation was made in the tunnel under Central Park.

The average ratio of molds to bacteria, as determined by the observations with filters, was 1 to 40 in the subway.

Effects of wind in the streets. The wind in the streets had a decided effect upon the numbers of bacteria collected from the air, both inside and outside of the subway. The averages show that five times as many bacteria were recoverable from the air in the streets with a wind of 18 miles per hour as with a wind of 9 miles.

Origin of the bacteria. No attempt was made to identify the different kinds of bacteria. To have undertaken to name the species, even with a great deal more time than was available and a special corps of bacteriologists, would probably have produced little result. Nevertheless, the conclusion was reached that most of the bacteria in the subway came from the streets. The principal reasons for holding this view follow:

1. The numbers of bacteria recovered from the air of the subway varied with the more decided changes in numbers in the streets.

2. At the subway stations the bacteria were more numerous near the stairways than at the remote ends of the platforms.

3. In the subway stations, the bacteria were more numerous on that side of the road into which the wind blew than on the opposite side.

4. There were more bacteria at the arrival ends of the platforms of the stations than at the departure ends.

5. Street dirt, probably containing large numbers of bacteria, was often carried down the stairways into the subway by inrushing currents of air and by the passengers.

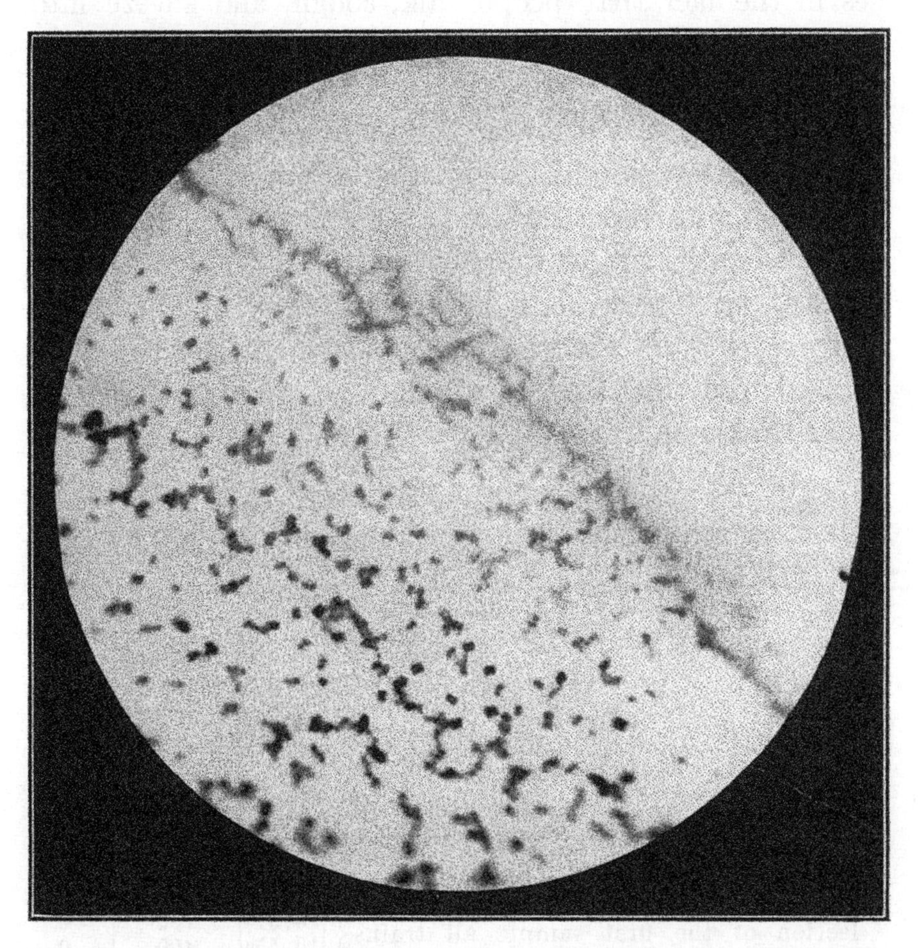

FIG. 36. Bacteria from the New York Subway magnified about 1000 times.

Although it seemed likely from these reasons that most of the bacteria in the air of the subway were derived from the streets, there was ground for concluding that some, and among them objectionable kinds of bacteria, were due to the presence of the people. (See Fig. 36.) It is practically certain when great crowds are packed together, as they often were in some stations and most cars, that dangerous bacteria are, at least occasionally, transmitted from person to person. An obvious feature of this danger lies in the fact that people talk, cough, and sneeze into one another's faces at extremely short range under such circumstances.

Effect of trains. The numbers of bacteria in the air of the subway varied with the amount of travel. They were most numerous when the trains were most numerous, and fewest when the trains were fewest.

When the trains were blocked many of the bacteria disappeared from the air. In one case the bacteria were reduced from 1800 to 250 in about an hour in this way. This is shown in Table VII.

TABLE VII.

EFFECT ON THE NUMBERS OF BACTERIA IN THE AIR OF THE SUBWAY PRODUCED BY A BLOCKADE LASTING AN HOUR, NOVEMBER 11, 1905

Place.	Time.	Microörganisms per cubic meter of air.	
		Bacteria.	Molds.
110th Street and Broadway station, north end, east platform. Soon after the collection of the first sample all trains stopped running. Finally, no passengers in subway	10.20a	1,300	0
	10.37a	750	0
	10.55a	400	0
	11.15a	250	0
Average	...	700	0

Effect of sweeping. The effect of sweeping the platforms with brooms, without first taking precautions against raising dust, was noted. On one occasion the numbers of bacteria were increased by sweeping from about 5000 to 13,000, and remained above 8000 for at least three-quarters of an hour — the time covered by the observation. The effect of sweeping is shown in Table VIII.

Effects on harmful microörganisms. It was not found that any harmful germs were capable of multiplying in the oil which dripped from the machinery of the cars upon the broken stone ballast and wooden ties of the roadbed.

The lubricating oil apparently removed and collected from the air large numbers of bacteria, many of which soon ceased to exist.

The pneumococcus was found capable of retaining its virulence in dried sputum in the subway for twenty-three days. This is in marked contrast to the findings of others, who have reported that the pneumococcus was killed in four hours in sunlight.

TABLE VIII

EFFECT ON THE NUMBERS OF BACTERIA IN THE AIR OF THE SUBWAY PRODUCED BY SWEEPING THE PLATFORMS IMPROPERLY

Place.	Time.	Microörganisms per cubic meter of air.		Ratio of bacteria to molds.
		Bacteria.	Molds.	
Fulton Street station, south end, west platform. Remote from openings to streets	10.25a	4,900	100	49:1
Porter began sweeping near by	10.41a	13,200	50	264:1
Still sweeping, but farther off	10.57a	8,100	0	...
Still sweeping, middle of platform	11.12a	8,500	0	...
Average	...	8,600	38	226:1

With few exceptions, there were not so many bacteria in the air of the toilet rooms as in the rest of the subway. In some cases the numbers were much greater.

Inefficiency of proprietary disinfectants. The proprietary disinfectants used in the toilet rooms had no germicidal or deodorizing value. Furthermore, they produced counter odors of a peculiarly unpleasant character.

Numbers of microörganisms in dust. The numbers of bacteria recovered from the dust of the subway averaged 500,000 per gram.

The largest number of bacteria found in subway dust was 2,000,000 per gram. Still greater numbers probably could have been found by selecting the specimens of dust toward this end.

For comparison with the numbers of bacteria found in dust from the subway, it is interesting to note that dust which had accumulated under similar circumstances in a Broadway theater contained 270,000 bacteria; in a new and fashionable hotel, 360,000; in a well-known Fifth Avenue church, 320,000; in the tallest office building in the city, 850,000; and in the attic of a country house one hundred and fifty years old, 110,000 bacteria per gram. The full results of these analyses are given in Table IX.

Dust which had accumulated in the subway contained over twice as many molds as dust collected in outside buildings. In the dusts the ratio of bacteria to molds was 89 to 1 for the subway, and 250 to 1 elsewhere.

ODORS

Odors were more or less prevalent at all times and at nearly all places in the subway. In some cases they were so faint as hardly to be noticeable, in others very decided.

The effects of the odors upon the passengers varied with

TABLE IX. — NUMBERS OF BACTERIA AND MOLDS FOUND IN ACCUMULATED DUSTS

Date, 1905.	Place.	Sample No.	Number of Microörganisms per gram of dust.		Ratio of bacteria to molds.
			Bacteria.	Molds.	
Oct. 23	96th Street station . .	1	550,000	1,500	367: 1.0
Oct. 24	Grand Central station .	2	1,000,000	2,500	400: 1.0
Oct. 25	14th Street station . .	3	2,000,000	500	4,000: 1.0
Oct. 26	Wall Street station . .	4	600,000	4,100	146: 1.0
Oct. 26	Brooklyn Bridge station	5	1,100,000	6,500	169: 1.0
Oct. 31	Columbia University station	6	81,000	1,600	51: 1.0
Nov. 1	Fulton Street station .	7	1,000,000	3,500	286: 1.0
Nov. 2	South Ferry station . .	8	300,000	5,600	54: 1.0
Nov. 3	Times Square station .	9	160,000	6,400	24: 1.0
Nov. 4	72d Street station . . .	10	600,000	2,900	07: 1.0
Nov. 6	66th Street station . .	11	460,000	0	...
Nov. 8	125th Street and Lenox Avenue station . . .	12	650,000	6,500	100: 1.0
Nov. 9	Mott Avenue station .	13	470,000	11,000	43: 1.0
Nov. 10	149th Street and Third Avenue station . . .	14	440,000	8,700	51: 1.0
Nov. 11	110th Street and Broadway station	15	290,000	6,000	48: 1.0
Nov. 13	Columbus Circle station	16	190,000	1,100	173: 1.0
Nov. 14	23d Street station . . .	17	160,000	2,300	70: 1.0
Nov. 15	Grand Central station .	18	650,000	550	1,182: 1.0
Nov. 16	14th Street station . .	19	500,000	17,000	29: 1.0
Nov. 17	Brooklyn Bridge station	20	370,000	11,000	34: 1.0
Nov. 18	Wall Street station . .	21	270,000	12,000	23: 1.0
Nov. 20	50th Street station . .	22	150,000	11,000	14: 1.0
Nov. 21	South Ferry station . .	23	120,000	4,200	29: 1.0
Nov. 22	96th Street station . .	24	340,000	5,000	68: 1.0
Nov. 23	Brooklyn Bridge station	25	800,000	8,300	96: 1.0
Nov. 24	Times Square station .	26	370,000	2,200	168: 1.0
Nov. 25	Grand Central station .	27	650,000	2,300	283: 1.0
Nov. 27	14th Street station . .	28	300,000	14,500	21: 1.0
Dec. 5	South Ferry station . .	31	51,000	3,900	13: 1.0
Dec. 6	Canal Street station . .	34	230,000	6,500	37: 1.0
Dec. 4	Attic of house 150 years old, Hadley, Mass. .	29	120,000	6,300	19: 1.0
Dec. 4	Attic of house 150 years old, Hadley, Mass. .	32	110,000	5,600	20: 1.0
Dec. 4	Cellar of old Hadley house	33	52,000	120,000	1: 2.3
Dec. 7	A fashionable hotel . .	35	360,000	4,400	82: 1.0
Dec. 7	A Fifth Avenue church	36	320,000	5,800	35: 1.0
Dec. 8	A Broadway theatre .	37	270,000	0	...
Dec. 9	A downtown restaurant	38	1,200,000	0	...
Dec. 9	Twentieth floor of an office building	39	850,000	1,500	567: 1.0
Dec. 9	Public ward of hospital	40	600,000	2,600	231: 1.0
Average of thirty subway observations			500,000	5,600	89: 1.0
Average of six New York buildings .			600,000	2,400	250: 1.0

the sensitiveness of the individual. To some persons the odors were exceedingly offensive, to others they were barely noticeable; many passengers soon became used to the odors and did not seriously object to them.

To persons unaccustomed to the subway the odors were unpleasant, and suggested that conditions existed which were injurious to health.

The odors were most apparent during hot, damp weather, at places where the greatest crowding occurred and where the least amount of ventilation took place.

Odors were far more often offensive in the cars than elsewhere, especially in the fall and winter months, when the windows were closed and the number of passengers was unusually large.

Methods of investigation. An effort was made to ascertain the main causes of the odors. It was not possible to look for them chemically or to measure them by other means than the senses, although samples of subway dust and air, when brought to the laboratory, often smelt unmistakably of the subway. By inspections in the subway and repair shops, by examining in the laboratory a large number of solid and liquid substances taken from the subway, and by attempting to duplicate the odors in closed chambers under different conditions of temperature and humidity, some of the causes of the odors were discovered.

Results of investigating the causes. The following conclusions were, in the author's view, justified by these studies:

Stone ballast. The stone ballast of the roadbed was responsible for part of the odor. This stone was made of broken trap rock, and its peculiarly slaty odor in the warm atmosphere of the subway was unmistakable. It could be most easily distinguished, especially at the more open stations, on damp days.

Frequently the odor of the trap was masked by other odors.

Lubricants. The oil used in lubricating the wheels and machinery of the cars was one of the principal causes of odor. Large quantities of this oil were allowed to drip from the machinery upon the ballast and ties of the road-bed when the subway was first put in operation.

Samples of the oil were obtained for experiment. It was not feasible to determine by analysis its exact composition, but in other ways it was ascertained that it was composed chiefly of petroleum and fish oil.

The quantity of oil used in the subway in the first year of operation appears to have been larger than had ever been used on an equal length of road. It amounted to over 200 gallons per mile of road per month.

In addition to this oil, about 150 pounds of gear grease were used per mile per month.

Much of the oil and grease was heated on the bearings, and some of it was volatilized. The car journals, motor armature bearings, and motor axle bearings were sometimes raised to a temperature of from 100 to 170 degrees Fahrenheit.

That the oil was distributed through the atmosphere of the subway was fully demonstrated. It was recovered from the dust by extraction with ether to the extent of 1.18 per cent by weight of dust.

Results of evaporation tests of the lubricating oil, according to the method of Gill, are given in the tables on following page.

LOSS IN WEIGHT OF SUBWAY OIL BY EVAPORATION — AIR TEMPERATURE

Sample No.	Per cent loss after twenty-four hours at 25–29.9° C.
1	0.31
2	0.35
3	0.45
Average	0.37

LOSS IN WEIGHT OF SUBWAY OIL BY EVAPORATION — TEMPERATURE OF PARTS LUBRICATED

	Wheel Journals.	Axle Bearings.
Sample No.	Per cent loss after eight hours at 74–76° C.	Per cent loss after eight hours at 110–115° C.
1	...	6.93
2	1.57	8.04
3	1.08	7.65
4	1.86	7.77
5	1.86	9.22
Average	1.59	7.92

It will be noted that at air temperature there was a loss of 0.37 per cent and 7.92 per cent at 110 and 115 degrees Centigrade. Under similar conditions whale oil, high grade fish and rape seed oils showed gains of from 1.01 to 5.39 per cent at 74 and 76 degrees, and losses of 0 to 2.65 per cent at 110 and 115 degrees, respectively.

Motors. Odors were given off by the hot motors acting upon various more or less volatile substances other than oil and grease. Among these substances were the insulating material covering some of the electric wiring and the paint upon the motor cases.

Electric sparking produced the odors of ozone and nitrous oxide.

The hot brake shoes gave off a peculiar odor.

Disinfectants. A pungent and unpleasant odor was produced by the proprietary disinfectants used in the toilet rooms. This odor was so penetrating that it was occasionally noticeable on the streets outside of the subway.

Tile cement. A strong and disagreeable odor was caused by an oily cement used in fastening decorative tiles in place at some of the stations. An ingredient of this cement was a cheap grade of fish oil. In order to disguise the fishy odor, creosote was freely mixed with the oil before mixing it with the cement. The result of these intermingled odors was peculiarly unpleasant. Fortunately, the odor of the cement, although very powerful at first, rapidly disappeared.

Hot boxes. Hot boxes, of which there were a considerable number when the road was first put in operation, at times produced a persistent and suffocating odor. Wool waste was used in packing the car journals, and when this caught fire its unpleasant smell could be distinguished through the subway for a long time.

Fuses. Occasionally a fuse was blown out and its odor was distributed up and down the line. When a fire occurred, as happened on a few occasions, the odor of smoke persisted in the part of the subway where the fire occurred for a surprisingly long period of time. In one case the odor was distinctly noticeable to passengers, as the cars passed the spot, three months after the fire had taken place.

Tobacco smoke. The odor of tobacco smoke was not uncommon at the subway stations. Rules existed against smoking in the subway, but they were not enforced. Lighted cigars, cigarettes, and pipes were occasionally carried even into the cars.

Concrete and fresh paint. Odors from new concrete and fresh paint were often noticed. The former was persistent, the latter transient.

Odors of human origin. Odors of human origin were sometimes present, but almost always close to people. They were most common during warm, damp weather and where there was much crowding. These odors often came from the clothing of the passengers. It was sometimes possible to learn the occupation of a workman by the odor of his clothes. Odors of coffee, garlic, bad teeth, liquor, cheese, and perfumery were some of the personal odors noticed.

The peculiar odor given off by clothing which had been hung in a kitchen was frequently noticed.

In fact, under the conditions of crowding, amounting frequently to close personal contact, it seemed that odors of practically every character connected with human existence were noticeable.

Excepting in rare instances, where ignorant employees were not kept under as strict supervision as their defective sense of decency required, the odors which permeated the general air of the subway did not point to conditions dangerous to health. Personal odors were detectible only at short range. When people are crowded so closely together that their breath and other body odors are offensive, there is always danger that disease may be transmitted from one to another.

The toilet rooms were much neglected at the time of this investigation, and often gave rise to an unpleasant local odor.

DUST

The dust of the subway was made the subject of study because of its unpleasant features and the possibility that it might play a part in producing or aggravating respiratory

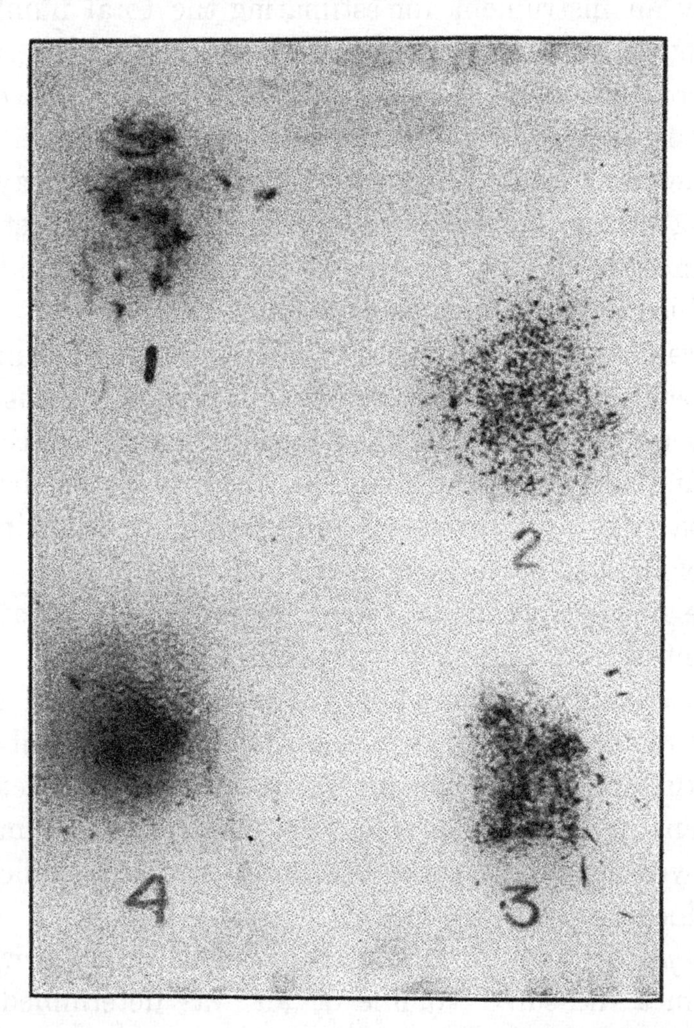

FIG. 37. Physical appearance of dusts. 1. From a fashionable hotel. 2. From the street (Broadway). 3. From a popular theatre. 4. From the subway.

diseases. Its possibilities for harm were to lie perhaps in its bacterial and physical composition.

Methods of examination. The dust was examined microscopically, chemically, and bacteriologically, by a special method which was devised for determining the gross weight of dust in a measured volume of the air, and by an instrument for estimating the total number of floating particles present.

Microscopical and chemical examination. The microscopical analyses were intended to show the shape of the particles and what could be ascertained in this way concerning their physical composition. The dust was examined under low powers of the microscope and with magnifications as high as 1200 diameters.

It was possible, by means of a common horseshoe magnet held beneath a piece of paper sprinkled with the dust, and slowly moved from side to side, to distinguish particles of iron and steel. These metal particles could be made to rise on edge and reverse their position by changing the pole of the magnet presented to them.

The chemical analyses were intended to indicate the amount of iron, organic matter, silica, and oil.

Bacteriological examination. The bacteriological analyses were intended to give some idea of the numbers of bacteria and molds present in dusts which collected at different points. (Fig. 43.) The bacteriological method employed in this work has already been sufficiently explained.

Weight of dust in air. At first the gross weight of dust in a measured volume of air was determined with a sugar filter, through which air was exhausted by means of an air pump. The amount of air which it was desirable to pass through the filter proved to be too great for an air pump of ordinary capacity. After experimenting with nearly every pump and blower on the market which promised to serve the purpose, a small Root's blower was employed, the blower being operated, as an exhaust, by

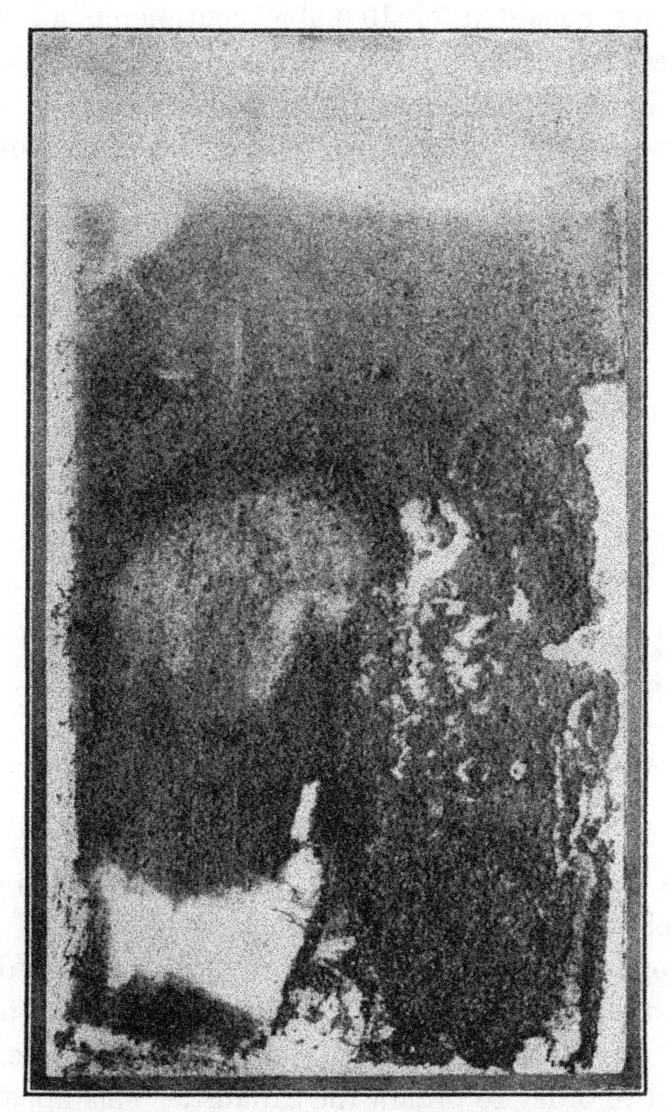

FIG. 38. Dust collected on a white tile exposed at the 59th Street subway station for one week.

hand, by means of a crank. A test meter, manufactured by the American Meter Company, New York, was used to measure the air. The meter was examined and found to be accurate to within $1\frac{1}{2}$ per cent when used in this way.

The filter consisted of 10 cubic centimeters of finely granulated sugar, contained in a glass funnel of 2.5 centimeters diameter, with straight upper sides. The sugar rested upon a plug of wire gauze and was 5 centimeters deep.

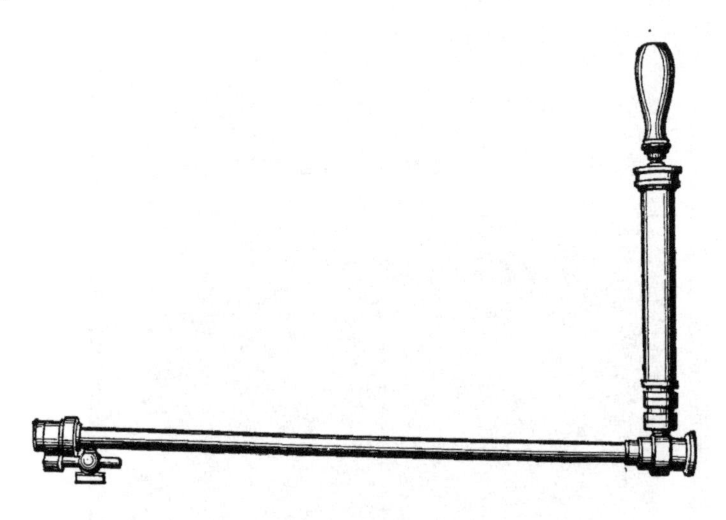

Fig. 39. Aitken's koniscope to determine the number of ultimate dust particles in air.

It was customary to pump 50 cubic feet, or 1416 cubic meters, of air through a filter for each observation.

The apparatus was so connected that the air passed first through the filter, next through the meter, and finally through the Root's blower, which was, of course, run backward in order to obtain the exhaust. This apparatus is shown in Fig. 40.

When the filter reached the laboratory, the sugar was carefully emptied into a beaker of distilled water. After the sugar was dissolved, the dust particles which remained were collected in a weighed Gooch filter containing a felt of asbestos. The filter was then washed with distilled water, dried at a temperature of 100 degrees Centigrade, cooled, and again weighed. The increase in weight was taken to be the weight of the dust. From the data so obtained, the weight of dust in milligrams per cubic meter of air was calculated.

Ultimate number of dust particles. The number of ultimate particles of dust was estimated by means of a koniscope, the invention of Professor John Aitken, F.R.S. A portable form of this instrument was imported from Glasgow for the purpose.

The koniscope has not been so much used in sanitary investigations as its merits deserve, and a few words may be desirable concerning it. It consists essentially of a brass tube closed at both ends by glass disks. Attached to the tube, near one end, is an air pump with suitable connections and stopcocks. (See Fig. 39.)

To estimate the number of particles of dust, the instrument was taken to the place where the atmosphere was to be examined. Air was pumped freely through the tube. The stopcock connecting the tube with the outside air was then closed. A rapid stroke of the pump made a partial vacuum in the tube, and this rarification produced a cloud or fog which could be distinctly seen by pointing the tube toward the light and looking through one of the glass disks at the ends. The dust particles served as nuclei about which the moisture condensed and so formed the fog.

The depth, color, and intensity of the fog indicated the number of dust particles present. Professor Aitken has invented a more elegant and exact, but less portable, dust counter with which the koniscope can be standardized.

This more refined apparatus was tried in the subway, but without wholly satisfactory results. It was very delicate in respect to adjustment, and required a better light than

Fig. 40. Apparatus for determining the weight of dust in a measured volume of air. To the right is an exhaust blower operated by hand. A gas meter is on the left. The air passes through a filter of sugar shown on top of the gas meter.

was obtainable. The approximate number of particles was usually estimated with the koniscope from the appearance of the fog, and in accordance with the table kindly supplied by Professor Aitken.

Results. The studies of dusts led to the following conclusions:

Physical character of the dust. In appearance, the dust was always black and very finely powdered. It was easily distinguishable by the eye from dusts collected in the streets, and in theaters, churches, office buildings, and mercantile and manufacturing establishments. (See Fig. 37.)

The subway dust had a peculiarly adhesive character, which caused it to attach itself securely to all surfaces, even when these were vertically placed and glazed. Dust collected on a white tile exposed at the 59th Street subway station is shown in Fig. 38. All parts of the subway which had not recently been cleaned and painted, or were not of a dark color, were sprinkled with this black dust when the investigation began.

The dust had a marked capacity for soiling linen and other articles of clothing. Straw hats and the light-colored garments worn by passengers of both sexes in summer were likely to be soiled by coming in contact with even small accumulations of the dust.

When examined microscopically, the dust was found to be composed of particles of many substances, conspicuous among which were fine, flat plates of iron. In fact, these iron particles could often be seen with the naked eye, glistening upon the hats and garments of persons who had been riding in the subway.

Particles two millimeters long were, on one occasion, taken from a magnet which had been carried in the hand on a ride of twenty minutes in the cars. By comparison, it was found that magnets hung up in the subway collected more particles of iron than magnets of the same size and strength hung up in an iron foundry or a dry grinding and polishing

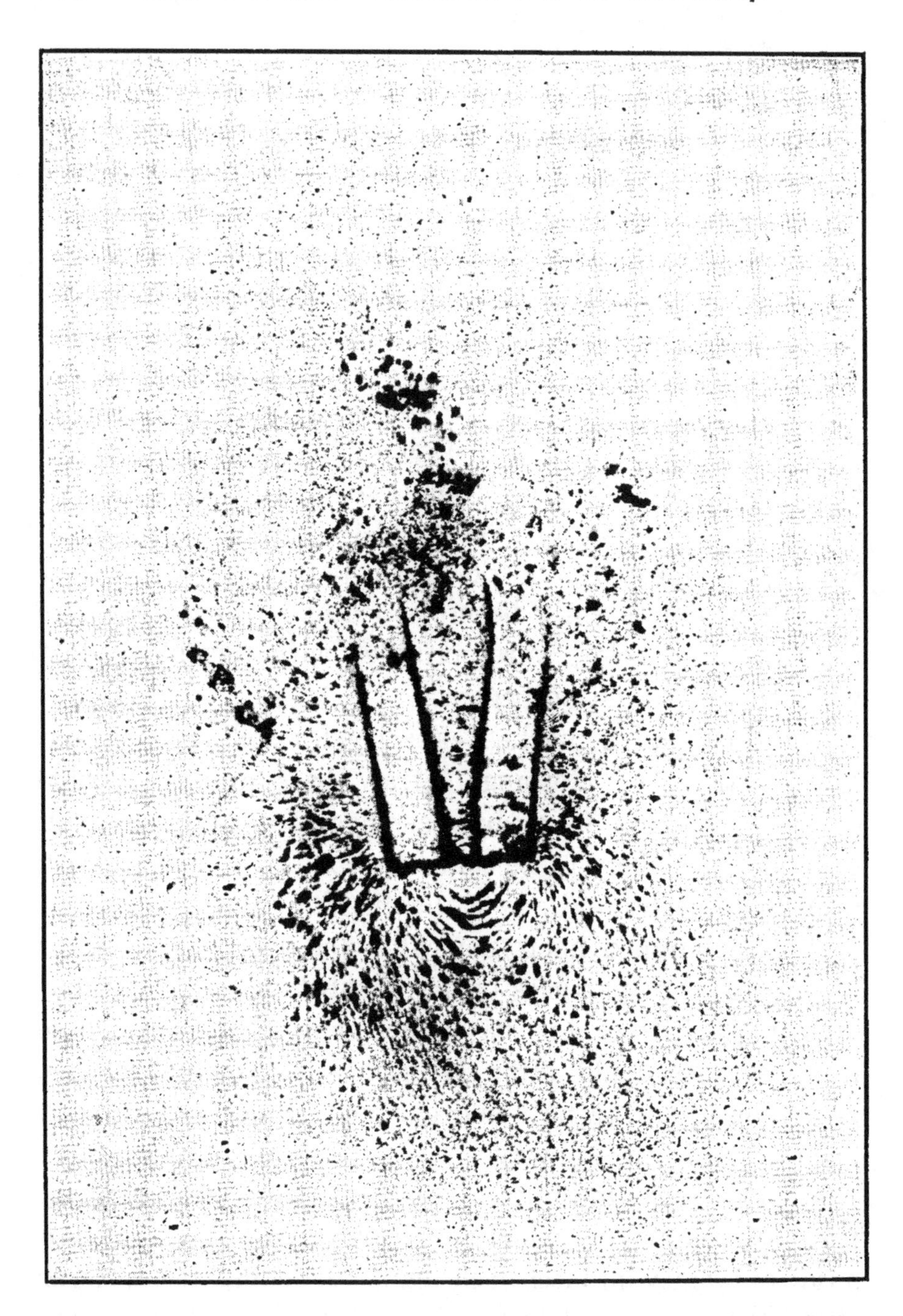

Fig. 41. Magnetic field formed by subway dust. A piece of paper was laid on top of a common horse shoe magnet and subway dust was sprinkled on the paper.

establishment. Figure 41 shows a magnetic field formed by subway dust.

The size, as well as the number, of the particles depended upon the place where they were found.

Many were so small that they floated in the air as dust. These generally escaped notice, except where beams of sunlight entered the subway or where the subway air emerged from some small opening into the sunlight in the streets, under which circumstances they glistened plainly.

Particles of subway dust, not iron, comprised bits of silica, cement, stone, fibers of wood, wool and cotton, molds, and undistinguishable fragments of refuse of many kinds.

Besides the dust which resulted from the grinding of metals, it was evident that the gradual wear and tear of many substances in the subway contributed to the dust.

TABLE X

RESULTS OF CHEMICAL ANALYSES OF ELEVEN SAMPLES OF ACCUMULATED DUST FROM THE SUBWAY

Date, 1905.	Place.	Total iron.	Silica, etc., insoluble in acids.	Oil.	Volatile and organic matter.
		Per cent.	Per cent.	Per cent.	Per cent.
Aug. 3	96th Street station	63.07	12.79	0.88	23.26
Aug. 3	14th Street station	41.77	26.39	1.43	30.41
Aug. 14	Grand Central station . .	67.35	12.65	1.23	18.77
Aug. 18	23d Street station	54.36	20.50	0.99	24.15
Aug. 17	Brooklyn Bridge station .	45.72	21.79	1.97	30.52
Aug. 21	33d Street station	69.66	12.34	0.91	17.09
Aug. 21	Canal Street station . . .	74.78	9.46	0.80	14.96
Sept. 19	116th Street and Lenox Avenue station.	66.69	13.84	0.96	18.51
Sept. 19	Times Square station . . .	68.42	7.45	1.00	23.13
Sept. 20	18th Street station	59.84	17.94	1.43	20.79
Sept. 20	28th Street station	62.58	16.28	1.42	19.72
	Average	61.30	15.58	1.18	21.94

Chemical composition of the dust. The separate chemical analyses of eleven samples of accumulated dust from the subway showed the following average percentage composition: total iron, 61.30, including 59.89 metallic iron; silica, etc., 15.58; oil, 1.18; organic matter, 21.94, as shown in Fig. 42. The results in full are given in Table X.

Origin of metallic dust. A large part of the metallic iron came from the wear of the brake shoes upon the steel rims of the wheels of the cars.

The wear upon the brake shoes was very severe. By weighing them when they were new and after they were worn out, and determining the number used, it was calculated by the operating company that 1 ton of brake shoes was ground up every month for each mile of subway.

The brake shoes consisted of cast iron with steel inserts.

There was also some loss to the rails and rims of the wheels and to the contact shoes which ran upon the third rail. Probably 25 tons per month would be a low estimate of the weight of iron and steel ground up in the whole subway every month.

Weight of dust in subway and street air compared. The average weight of dust found in the subway by the use of the sugar filters, using all of the results, was 61.6 milligrams per thousand cubic feet of air, or 2.25 milligrams per cubic meter; in the streets, 52.1 milligrams per thousand cubic feet, or 1.83 milligrams per cubic meter; difference, 9.5 milligrams. The maximum amount found in the subway was 204 milligrams.

Twenty-three comparative tests were made to determine with particular care the weight of dust per thousand cubic feet of air inside of the subway and in the streets at the same time and as near the same place as possible.

These showed an excess of dust in the subway of 47 per cent over that outside. In five cases there was more dust outside, the greatest excess being 30 per cent. In the other eighteen cases the excess of subway dust over street dust ranged from 11 to 800 per cent.

Weight of dust inhaled by passengers. The weight of dust which the average passenger inhaled in one-half hour

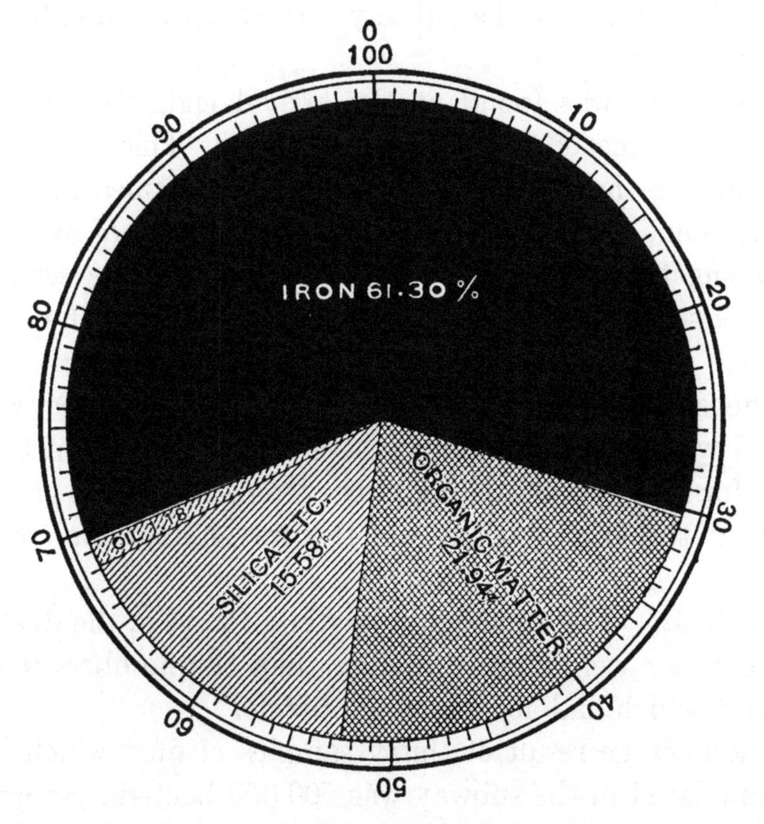

Fig. 42. Composition of subway dust as determined by chemical analysis of eleven samples.

in the subway was very slight. Assuming that 360 cubic centimeters, or 22 cubic inches, of air were taken in at each breath and that the passenger breathed eighteen times a minute, the total quantity of air which passed into the lungs in half an hour was about 6.88 cubic feet, or 6.50 cubic meters. Using the average of all results, or 61.6 milligrams per thousand cubic feet, as the weight of dust suspended in the atmosphere, it appears that the average passenger took into his nose or mouth .42 milligrams of dust in a ride of half an hour.

Variations in the amount of dust. The amount of dust found in the air of the subway varied with a number of circumstances.

More dust was found at the arrival ends than at the departure ends of the station platforms. This was probably due to the fact that the brakes were applied near the arrival ends, and to the fact that the currents of air from incoming trains helped to carry dust from those sections of the subway which lay between stations to the platforms.

The stations where the greatest weights of dust were found were express stations; there the amount of metallic dust formed by the braking of the trains was much greater than at the local stations and the travel from the streets greatest.

Bacteria. The numbers of bacteria found in the dust of the subway were usually smaller than the numbers found in dust which had accumulated outside.

The average result of thirty samples of dust which had accumulated in the subway was 500,000 bacteria per gram of dust. The average obtained from six samples of dust which had accumulated under what appeared to be comparable circumstances in different buildings in New York was 600,000.

The largest number of bacteria found in a sample of subway dust was 2,000,000.

CONCLUSIONS

A review of the results of the investigation so far warrants the following brief statement of the most essential facts determined with respect to the quality of the air.

According to usual sanitary standards, based on chemical and bacteriological analyses, the general air of the subway was always and everywhere satisfactory. The air in the cars is not included in this statement.

According to public opinion, based on the testimony of the senses, the air was everywhere unsatisfactory, especially during the summer months.

The author's own conclusion was that the general air, although disagreeable, was not actually harmful, except, possibly, for the presence of iron dust. The strong draughts in winter at the stations and the lack of sanitary care exercised over the subway were, however, worthy of careful consideration in this connection.

The high temperature of the subway was its most noticeably objectionable feature. Had it not been for the heat, it is probable that the other unpleasant features would have failed to arouse serious protest. The heat, as is well known, was due to the conversion of electric power into friction. The amount of heat given off by the passengers was so small by comparison as to have had practically nothing to do with elevating the general temperature.

The heat was most objectionable in the mornings and evenings of summer during the hours of greatest travel and when the air outside was cooler than during the rest of the day.

The heat did not indicate that the air was vitiated or stagnant, as was popularly supposed. The subway was hot because a great deal of heat was produced in it, and stored by the materials of which the subway was built.

That the heat did not escape rapidly enough for comfort was no proof that the air was not renewed often enough for health.

The carbon dioxide and oxygen analyses indicated that the products of respiration were rapidly carried away. Among the 2200 carbon dioxide determinations, most of which were made in the subway, no sample of air was taken which contained above 8.89 parts of CO_2 per ten thousand volumes, and this amount was found under circumstances which must be regarded as exceptional.

The average excess of carbon dioxide in the subway over that in the streets, 1.14 parts per ten thousand volumes, showed that the air was renewed with remarkable frequency. In the absence of a census giving the number of passengers in different parts of the subway at different hours, it was impossible to calculate just how frequently the air was renewed; but from such estimates as it was possible to make it seemed not improbable that the air of the whole subway was completely renewed at least every half hour.

It is true that the renewal occurred somewhat more frequently in some parts of the subway than in others, but the exchange was always and everywhere abundant. We must except, of course, from this statement, the cars when crowded and closed, and other places where dense crowding occurred.

The controlling condition which regulated the extent to which the air was renewed was the freedom with which it was forced in and out of the subway. The air was best where the subway was most open to the streets, and, conversely, it was least satisfactory where the subway was most enclosed. More blow-holes would have greatly improved the conditions.

The movement of the trains set in motion the essential

ventilating currents. This they did, first, by forcing subway air out and bringing street air in at openings; and second, by moving the air through the subway between openings.

It was fully demonstrated that there were no pockets or other places where air stagnated. Diffusion was every-

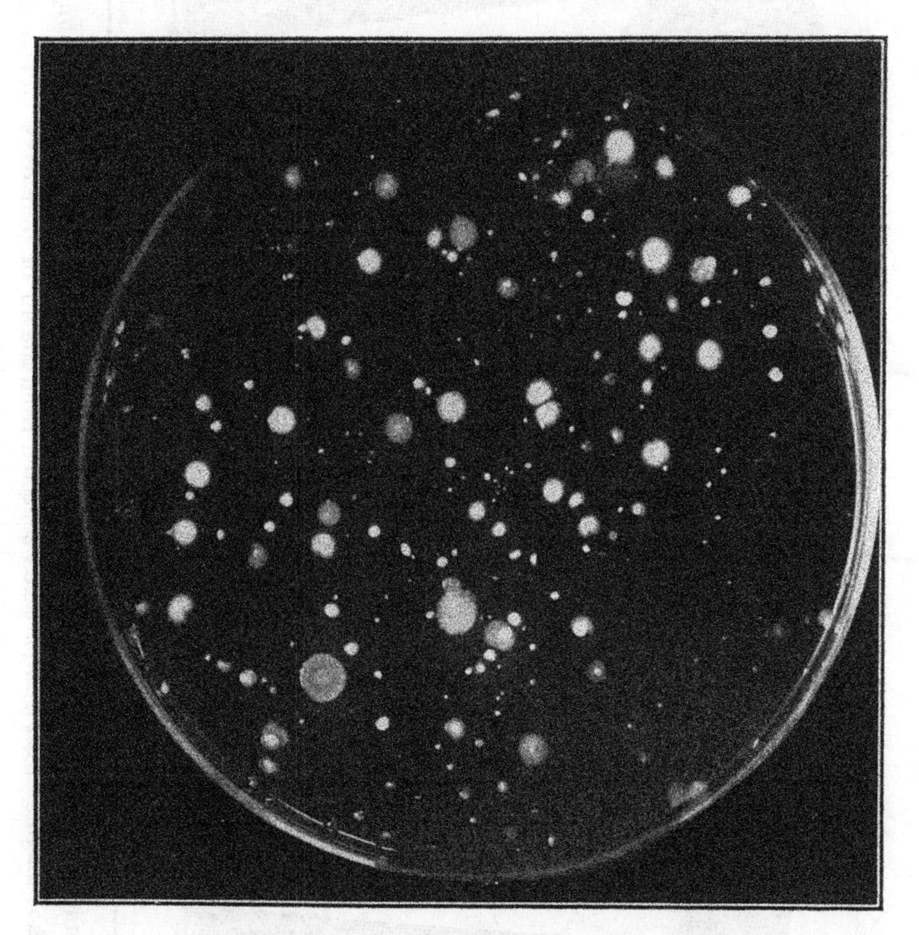

Fig. 43. Colonies of bacteria from the dust of the New York subway.

where rapid, complete, and satisfactory. The cars are excepted in these statements, as already indicated.

The fact that there were only about half as many bacteria found in the air of the subway as in the air of the streets

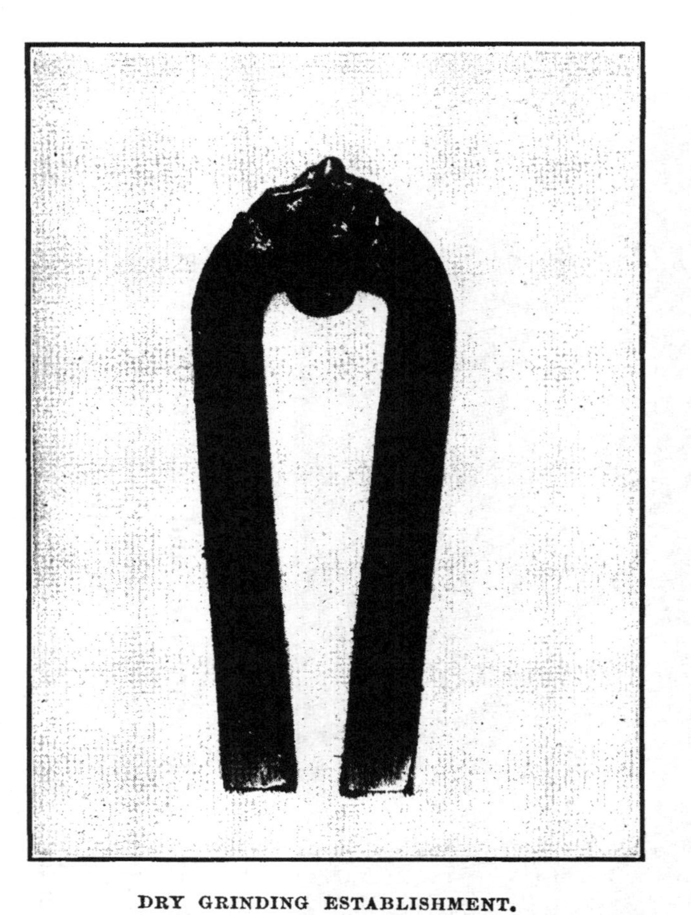

DRY GRINDING ESTABLISHMENT.

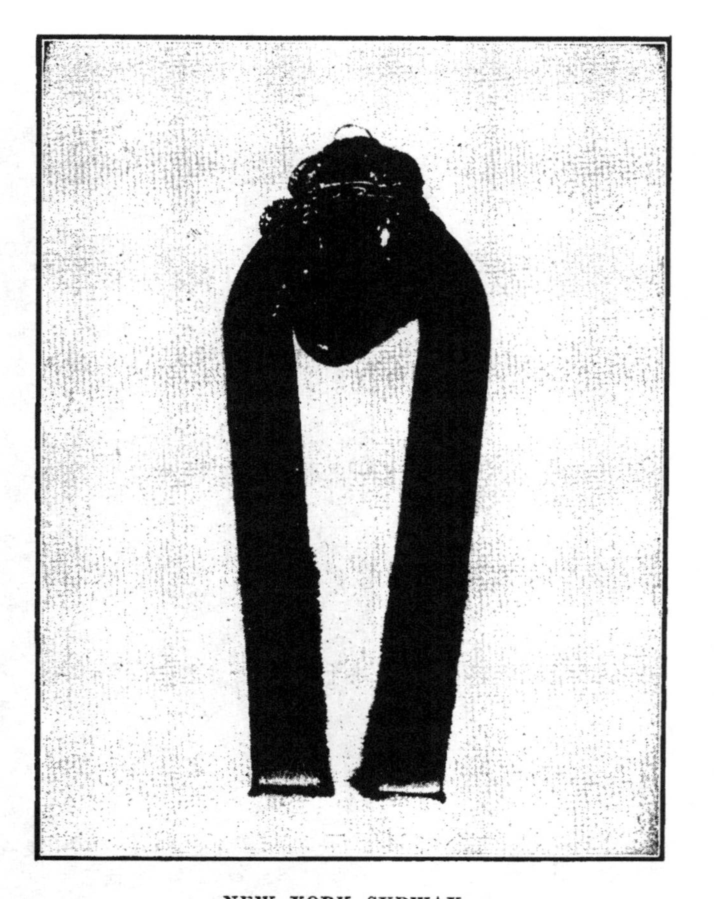

NEW YORK SUBWAY.

FIG. 44. Horse shoe magnets with iron particles collected from the air in four days. — The magnets were freely suspended by means of twine.

under which the subway ran gave ground for the opinion that the bacteriological condition of the subway air was satisfactory, although too much reliance should not be placed upon this guide to its condition. Judgment on this point would have been more conclusive had it been possible to demonstrate that no more harmful bacteria existed in the subway than in the air outside. This was beyond the practicable possibilities of bacteriological technique.

The odors of the subway, like the heat and dust, were objectionable, apparently, chiefly because they were disagreeable. They resulted largely from the operation of the trains. They were, in the author's opinion, to a large extent preventable.

The sanitary significance of the characteristic black dust of the subway, containing, as it did, over 61 per cent of metallic particles, remained to be considered at the close of this part of the investigation.

CHAPTER VII

HEALTH OF NEW YORK SUBWAY EMPLOYEES

This chapter is practically identical with a report made in May, 1907, to the Board of Rapid Transit Commissioners for the City of New York, and describes an investigation by the author into the possible effects of the metallic dust of the first New York subway on the health of the employees.

Thanks are due to many persons for help. The Interborough Rapid Transit Company, through Mr. Frank Hedley, General Manager, granted requests for information concerning the men and furnished the 100 employees who were examined.

The physical examinations and analyses were made with much skill by Dr. James Alexander Miller, Instructor in Physical Diagnosis at the College of Physicians and Surgeons, assisted by Doctors H. C. Hanscom, J. M. O'Connor and I. O. Woodruff. In the autopsies and subsequent histological examinations thanks are due to Dr. J. H. Larkin, Adjunct Professor of Pathological Anatomy, and to coroner's physicians Doctors T. D. Lehane and P. F. O'Hanlon. To Dr. Frank B. Mallory, Associate Professor of Pathology, Harvard Medical School, who collected records to show the pleurisy found in 1008 autopsies performed at the Boston City Hospital, the author is also indebted. Finally, a number of eminent pathologists and medical practitioners aided the work by valuable council, suggestions and opinions.

PLAN OF THE INVESTIGATION

It was intended that the investigation should be made so as to detect any physiological effects which might be caused by the dust of the subway, whatever they might be. Special care, however, was taken to look for early signs of more serious disease of the lungs, which exists to an excessive extent among persons engaged in dusty occupations. Sufficient time had not elapsed since the subway was opened to reveal the symptoms of pneumokoniosis had this condition existed, but it was hoped that if this disease was progressing some sign of it could be detected.

Physical examinations were made of a sufficient number of subway employees to determine the condition of the average man. Supplementary to these, bacteriological and chemical analyses were made of their sputum, urine and sweat.

To compare the physical condition of the subway men with that of persons not engaged in subway work, examinations were made of 200 men representing twenty different occupations.

To help arrive at an understanding of the possible effects of the subway dust, data referred to in the first part of this investigation were collated concerning its chemical composition, physical properties, and the weight of dust in a given volume of subway air. The bacteria associated with this dust were also considered.

The condition of the air and the work of the men were compared with the conditions which exist in such vocations as stonecutter, knife grinder, metal polisher, and other dusty occupations in which a high mortality occurs.

It being desirable to obtain an accurate understanding of the anatomical condition of the men, and as this could be had only by dissecting their bodies after death, arrangements were made to have as many autopsies as practicable performed upon the remains of employees killed by accident in the subway during the period covered by the investigation.

Light was thrown upon the frequency with which pleurisy was found at autopsy by reviewing the records of a large number of reports of post-mortem examinations made elsewhere.

THE CONDITION OF THE AIR

Investigations which had been made for the Board from July, 1905, to January, 1906, showed that the chemical condition of the air of the subway, in spite of unpleasant odors and heat, was remarkably good.

The carbon dioxide, as determined by 2084 analyses covering practically all times and places, was found to be but little higher in the subway than in the streets: the average for the subway was 4.81, and for the streets 3.67. These figures represent parts of carbon dioxide in 10,000 volumes of air.

There was ample oxygen. The average of eighty analyses gave 20.60 per cent of oxygen for the subway as against 20.71 per cent for the streets.

The analyses, checked by observations of air currents, indicated that the atmosphere of the subway was completely renewed at least every half hour before any material change had been made in the methods of ventilation.

Summarizing the opinions which the author formed at the conclusion of that investigation, the principal possibilities for harm in the air, aside from rapid changes of temperature and strong draughts, lay in the presence of the black metallic dust. The principal characteristics of this dust will now be described.

Physical and Chemical Composition of the Dust. When examined microscopically the dust was found to be composed of particles of many substances, including innumerable fine, flat plates of iron. These iron particles could be seen by a sharp eye glistening upon the hats and garments of persons after a short ride in the subway. The clothing of the employees gathered this dust, and their hands, bodies, and linen became discolored with it.

Large particles of iron could readily be seen at the stations, upon the roadbed. A common horseshoe magnet suspended at the breathing line would, in a few days, collect a surprisingly large amount of iron dust. Of two magnets hung up — one at the Grand Central Station of the subway, and another in a dry-grinding establishment, the subway magnet collected by far the most dust. (See Fig. 44.)

Eleven samples of subway dust were analyzed chemically, with results given in Table X, page 185.

A glance at this table shows that the average amount of iron in the dust was 61.3 per cent. The samples were collected from smooth, clean surfaces upon which the dust had been allowed to settle from the air.

In addition to the iron particles, the dust contained bits of silica, cement, stone, fibers of wood, wool, cotton, silk and other textile materials, molds, and indistinguishable fragments of refuse of many kinds, resulting from the wear and tear of the subway and clothing of the passengers. In fact, everything in the subway susceptible of wear contributed to the dust. In addition, refuse from the streets was carried into the subway by inflowing currents of air and by passengers.

Bacterial composition of the air and dust. On the whole, the numbers of bacteria found in the dust of the subway were smaller than the numbers found in dust from the streets. The average obtained on analyzing thirty samples of dust from the subway was 500,000 per gram of dust. The average number found on analyzing 141 samples of subway air was 3200 per cubic meter. These figures were about one-half as large as were found for the streets.

These numbers represent bacteria capable of growing on beef-extract-agar, of $\frac{1}{2}$ to 1 per cent acid reaction to phenolphthalein, at the temperature of the body and at such a rate that colonies could be counted at the end of forty-eight hours.

There was reason for believing that some of the bacteria in the subway were more harmful than those generally found outside. The absence of sunlight in the subway prolonged the life of the germs of some diseases. The pneumococcus, believed to be the cause of lobar pneumonia, was found by experiment to be capable of living twenty-one days in the subway as against four days in the streets.

The lack of enforcement of the city ordinance against spitting and the frequency with which passengers and employees expectorated upon the tracks, platforms, and stairways, increased the danger from tuberculosis and other respiratory diseases. The excessive crowding exposed the employees, particularly those holding the grades of guard and conductor, to still greater danger of infection.

Sources of the iron dust. A large part of the metallic dust came from the wear of the brake shoes upon the steel rims of the wheels under the cars. It was calculated that one ton of brake shoes was ground up on every mile of the subway every month. In addition, there was some loss of metal from the rails, especially at the curves. So great was this wear that an especially durable steel was at length made to withstand it.

The rims of the wheels and the contact shoes which supplied the motors under the cars with electricity from the third rail contributed some weight of metal to the dust. Probably twenty-five tons would be a low estimate of the total weight of iron and steel ground up in the twenty-one miles of subway every month.

It must not be supposed that all of this great amount of ground iron floated in the air. Some of the pieces were so large that they fell immediately to the track and remained there. Others were raised only for brief moments by violent eddies produced by the trains. Large quantities were caught by the ties and broken stone ballast, which were continuously sprinkled with lubricating oil from the trains.

Many of the particles were so greasy that they adhered firmly to whatever surfaces with which they happened to come in contact. The smallest and probably the freshest particles remained longest in the air. It was these which constituted most of the dust used in the analyses. It was these which were breathed.

Some dust was carried up into the streets by air currents which were forced out through the station stairways and blow-holes by the trains. The trains also kept some dust in suspension in the subway. Had there been no trains the dust would have quickly settled from the air. This is shown by the bacteriological experiment recorded in Table VII, page 168, in which the bacteria acted like exceedingly minute dust particles.

Weight of dust in air. The average weight of dust in subway air was found to be 61.6 milligrams per thousand cubic feet of air, or 2.25 milligrams per cubic meter. This was somewhat more than was found in the streets under parallel conditions. The figures for the streets were 52.1 milligrams per thousand cubic feet, or 1.83 milligrams per cubic meter. The average for the subway was made up of the results of 146 analyses made at points and at times especially selected to give a correct knowledge of the normal conditions.

These analyses show that the total amount of dust in all the air contained in the subway at any time from the 96th Street station to the Brooklyn bridge was 3½ pounds. The dust was not quite evenly distributed through the air. There was more at express stations than elsewhere. At any given station there was more dust at the arrival ends of the platforms than at the departure ends. These differences were, however, slight.

Weight of dust inhaled. The weight of dust which an employee took into his mouth or nose during the course of a day of ten hours could be computed from the results of the analysis just referred to. Assuming that 360 centimeters or 22 cubic inches of air were taken in at each breath, and that the employee breathed at the average rate of eighteen times per minute, the total quantity of air which passed into his lungs in ten hours was 6.86 cubic feet, or .19425 meter. Taking 61.6 milligrams per thousand cubic feet as the weight of dust suspended in the atmosphere, it is found by calculation that an employee took into his nose or mouth 8.4 milligrams of dust in ten hours. This is 3066 milligrams per year, or forty-six grains. It will be shown presently that only a small part of this could get into the lungs.

Reliable data are lacking to show the weight of dust which exists in the air of steel-grinding and other establishments where pneumonokoniosis is produced by dust. Hesse found, some years ago, from 72 to 100 milligrams of dust in a cubic meter of air in an iron foundry, and 14 milligrams per cubic meter in the air of an iron mine.

The dangers of the dust. The possibility that the dust might cause injury to the eyes, to the skin, and to the respiratory apparatus was considered in this investigation; but the condition of the throats and lungs of the employees received the largest share of attention.

Injurious properties of subway dust. Inasmuch as dust may do harm in many ways, it may be well to describe briefly how the subway dust was regarded in its relation to the health of the subway employees.

Chemical composition. There was nothing about the chemical composition of iron particles to make them especially dangerous. Iron is not like lead and other poisonous substances in this respect. If the subway dust had been composed of silica, probably its action would have been no different.

Mere quantity. The amount of dust breathed was not great enough to be injurious solely on account of its bulk. In this respect the atmosphere of the subway was wholly unlike that of flour mills and cement mills. There the quantity was vastly greater.

Bacteria. Dust when breathed may cause disease by carrying bacteria into the throat and lungs. This was a matter worthy of some attention, in view of bacteriological conditions in the subway already described. Particles of dust which carry or accompany harmful bacteria are among the most injurious kinds of dust.

Mechanical or physical composition. Dusts whose consistency most resembles that of the organs which they invade are least harmful, so far as physical composition is concerned, and the more jagged in outline and resistant in texture they are, the greater is the capacity of the particles to do harm. They irritate the delicate organs with which they come in contact, and so open the way for the entrance of pathogenic microbes. The most injurious of all dusts are composed of such substances as iron and steel.

So far as the subway dust was examined, the mechanical and bacterial conditions were of most interest.

Contributing factors. Various factors predispose persons to respiratory dust diseases. Among these may be mentioned:

1. The existence of some respiratory disease already, as, for example, tuberculosis.

2. Predisposition to respiratory disease, whether this predisposition is inherited or constitutional, increases susceptibility.

3. A neurotic condition, in which the person anticipates or expects evil effects to follow the inhalation of a dusty atmosphere, increases the liability.

4. Exertion, requiring the breathing of unusually large quantities of air, bringing into the lungs more dust, must be recognized as a contributing factor. Mouth breathing may be included in this category.

5. Humid air, draughts, or an atmosphere in which rapid changes of temperature occur, contribute to the evil possibilities of dust.

6. An especially severe use of the voice is unfavorable.

Of all these factors, the amount of air breathed, the atmospheric changes, and the severe use of the voice seemed to be especially worthy of consideration.

Natural defenses against dust. In normal health the delicate structure of the lungs is protected in various ways against the entrance of dust particles from the air:

1. The nose and throat are themselves effective barriers. Only a very small proportion of the dust particles which enter the mouth or nose escape the moist and irregular channels which lead to the throat.

2. If a particle passes the mouth or nose, it is almost certain to be arrested by the mucous membrane of the trachea and lower air passages. Here myriads of moving cilia carry it to a point from which it can be removed by the conscious mechanism of coughing.

3. If the particles go further they enter the bronchioles, and from these pass to the air cells of the lungs.

4. When particles of dust reach the air cells they do not necessarily pass into the tissues. They do so only when they penetrate the endothelium with which these cells are lined.

Minute foreign substances may, however, be taken into the tissues before reaching the air cells of the lungs, especially when the normal activity of the mucous membrane is reduced. This is not uncommon among persons who breathe a dusty atmosphere. When particles are absorbed, whether in the throat or lungs, they generally enter the lymphatics and are retained by the nodes, or filtering arrangements, with which the lymphatic system is provided. Only in rare instances do foreign particles penetrate through the lymphatic system to the blood.

Infection through pathogenic microbes occurs when the protective barriers peculiar to the surfaces of the delicate mucous lining of the air passages become injured and the natural resistance toward them is reduced. It is not improbably due largely to the constant irritation produced by dust upon the mucous membranes that respiratory diseases are so common among city dwellers.

Condition of throats and lungs of city dwellers. Respiratory diseases are extremely common among persons who live in cities, pneumonia being frequently recorded as the leading cause of death, with tuberculosis following closely. Bronchitis and laryngitis are equally common, though less fatal, and pharyngitis and rhinitis still more prevalent. The minor affections not infrequently lead to the more serious ones.

When examined after death, the lungs of city people can easily be distinguished from those of dwellers in the country, the bright, rosy color which is natural to the latter being changed to gray and sometimes black by particles of soot and dust which have gotten into them from the air.

Mingled with the dust in the lungs of all city dwellers are particles of iron. So great is the wear of iron, particularly from the wheels of vehicles, the brakes of street and elevated railway cars, and the shoes of horses, that it was impossible during this investigation to find a single specimen of dust in New York which did not contain particles of metallic iron. Iron particles were collected from the surface of fresh snow on Liberty Island in the center of New York Bay, a mile or more from the nearest land, thirteen days after an earlier snowstorm had covered the ground and kept dust from being blown from places where it had settled. White marble buildings in New York lose their original color within a year and become noticeably yellow. The amount of iron dust which gets into the lungs is extremely small, but it can be detected with certainty by the microscope and by analysis.

It seems unnecessary to refer to other dusty particles

which get into the lungs of city dwellers. The city streets are notoriously dusty. The dust consists of a pulverized mass of refuse in which building sand, ashes, and dry horse manure are conspicuous ingredients. The tendency of this dust is to settle to the earth, but excepting immediately after rain or snowstorms it never is absent from the air. No building in New York is high enough to escape it. It is most objectionable in the crowded streets. At a single breath a pedestrian may take into his nose or mouth a much greater quantity of dust than the average subway employee gets in a month.

Because of the peculiarly large amount of iron in the dust of the subway, the disease known as siderosis, which exists most commonly among metal polishers, knife-grinders, and others engaged in working in metal, was given special attention.

Disease due to iron dust. The inhalation of iron dust produces evil effects in three ways:

1. By diminishing the respiratory efficiency of the lungs through a loss in their elastic property.

2. By reducing the resistance of the organs to invasion by harmful bacteria.

3. By infecting the lungs through a transportation of disease germs to places favorable for their inoculation.

The earliest symptoms of siderosis are catarrh and bronchitis, but shortness of breath is pronounced by all authorities to be the most characteristic symptom. Eventually there follows what appears to be phthisis without the presence of tubercle bacilli. Yet genuine infective phthisis is the most common cause of death.

The cause of the unpleasant symptoms is sometimes not discovered until the exposure has been endured for years, depending upon the amount of dust in the air and the personal resistance to it. Even in fork-grinding, among the most dangerous of dusty occupations, the effects may be delayed for decades.

Probably a great many men engaged in dusty occupations pass their lives without suspecting the cause of the uncomfortable symptoms which they experience. There is no doubt that large numbers die from infectious pulmonary diseases who do not know that the breathing of dusty air has led to their infection.

A writer in a recent periodical[1] has shown the startlingly high rate of death among various classes of metal workers in America who are apparently in ignorance of the peculiar danger of their occupation.

The death rate among steel grinders and others at Solingen, Germany, for the ten years, 1885–95, is shown in the following table, in which the number of deaths from consumption is given in 1000 deaths from all causes in Germany.[2]

TABLE XI

DEATH RATES FROM PHTHISIS AT SOLINGEN, 1885–1895

Age.	Grinders.	All males.
14–20	25.8	40.0
21–30	84.4	69.9
31–40	75.9	47.0
41–50	79.3	36.0
Over 50	68.7	25.8

[1] *The Independent,* "A Story of the Death Claims." Andrew Hellthaler. December 27, 1906.

[2] Handbuch der Medizinischen Statistik. F. Prinzing. Jena, 1906, p. 489.

DESCRIPTION OF THE SUBWAY EMPLOYEES

Of the force of about 3000 men employed by the Interborough Rapid Transit Company to operate the subway, about 800 were motormen, conductors, or guards upon the trains, about 800 ticket sellers, ticket takers, or porters at the stations, and about 1400 switchmen, trackmen, mechanics, painters, engineers, or others engaged on the road, in the shops, power houses, or elsewhere. This investigation was restricted chiefly to the uniformed force, consisting of trainmen and station men.

All who occupied responsible positions with respect to the operation of the trains were examined physically by the company before they were employed, records being kept of their age, weight, height, respiratory capacity, sight, color sense, hearing, and heart action.

Absences from work for less than two weeks were not, as a rule, inquired into by the company, but in the event of serious sickness the men were examined medically before they were allowed to resume their work.

In this investigation the employees sent by the company for examination were, at the author's request, forty-five motormen, forty-five conductors, and ten switchmen, this list including employees who had been longest on the road and whose work had kept them most closely under conditions similar to those experienced by the traveling public.

Physical appearance of the men. Nearly all the men were of fine physique. Capacity to do hard manual labor was not demanded of the motormen, nor was it necessary that the conductors should have more than ordinary strength; but inasmuch as men who were eligible to these grades were recruited largely among persons who had had other railroad experience, it was to be expected that the physical standard would be high.

The men were all between twenty-one and forty-seven years of age. Their average height was 5 feet 8¼ inches, and their average weight 169 pounds. Their general appearance of health was excellent in fifty-two cases, good in thirty cases and poor in only two cases.

Sixty-nine per cent of the men claimed to be citizens of the United States. About half were city-bred.

Regular duties of the men. The motormen were from the regular force engaged in operating the trains. Their employment required them to sit in a small compartment at the forward end of the car at the head of a train, where they operated a number of small hand levers. The position and duties of these men prevented them from doing any physical work. They were on duty, at most, ten hours each day.

The conductors serve in the capacity of guards, with some additional duties and responsibilities not of interest in this investigation. They are stationed between the first and second cars of the trains. They stand while at work and are required to make considerable exertion in opening and closing the heavy doors of the cars at the stations. In calling out the names of the stations amid the noise of the moving trains, the throats of the conductors are put to considerable strain.

The duties of the switchmen require them to couple and uncouple cars and switch them back and forth from one track to another at the yards and storage places. This force is largely composed of men who are in the line of promotion to motormen. Of the ten switchmen examined four had been accustomed to railroading and two to indoor work exclusively.

Medical history before and after entering subway employment. The men were asked to give histories of themselves before and after entering upon their subway employment. In some cases the histories given were undoubtedly unreliable, and in a few cases the men were evidently disinclined to talk; but, in general, their attitude was one of frankness and honest coöperation, and it seemed safe to put considerable reliance upon their statements.

Forty per cent had a history of previous serious illness. Sickness had been divided proportionately between the three classes. The longest time lost through illness before going to work in the subway was five and a half months.

The average length of time that the men had worked in the subway up to the time of examination was 18.2 months. Only two men had been employed less than one year; these had been working ten months.

Thirty of the 100 men claimed to feel in better condition at the time of examination than when they first began to work in the subway; five felt in poorer health; sixty-five were unchanged. In fifty-four cases the weight had increased, this increase varying from $7\frac{3}{4}$ pounds to $15\frac{1}{2}$ pounds. In eighteen cases there had been a decrease.

There were fifteen cases of illness reported to have occurred during the period of subway service. Of these only nine were affections of the respiratory apparatus. There had been three cases of tonsilitis and one case of bronchitis. In twenty-seven cases time had been lost from illness, but most of these illnesses were apparently of a trifling character. The conductors had lost more time than motormen or switchmen.

Twenty-five men spoke of a metallic taste in the mouth, although this point was not mentioned by the others. In seventy-seven cases a decided and peculiar yellow stain was noted on the clothing where it was moistened by perspiration. In fifteen cases these yellow stains were observable on the body, on underclothing, and on bed-clothes even after bathing. Unusual drowsiness was mentioned in forty-six of the fifty cases inquired into.

In fifty-seven cases some sort of precaution, such as douching, was taken by the men to protect the nose and throat. In seventeen cases the nose and throat were douched every day.

RESULTS OF THE PHYSICAL EXAMINATIONS OF THE EMPLOYEES

About forty minutes were consumed in examining each man. The eyes, nose, and throat were first examined. The men were then required to strip to the waist and an examination was made of the organs of the thoracic and abdominal cavity. Measurements of the chest completed the examination. All observations and verbal information gathered from the men were noted in blank forms prepared for the purpose.

Examination of principal organs and eyes. In nearly all cases the principal organs were in good condition. The average pulse rate was eighty-five, the highest 104, and the lowest fifty-six. The cervical glands were enlarged in sixteen cases.

The eyes were found to be slightly red or irritated in 39 per cent of the men, but there was no congestion of the ocular conjunctiva. These conditions were no more prevalent in one class than in another.

Examination of the upper air passages. Abnormal conditions found in the nose and throat differed only in degree from those usually found in dwellers in cities. Bony irregularities favor catarrhal conditions, and these were probably responsible for a good many of the cases of catarrh noted in these examinations.

In sixty-eight cases rhinitis was found. It was marked in two conductors and two motormen. Bony abnormalities of the nose existed in forty-five cases.

Forty-three of the men gave a previous history of catarrh. Fifty-four had catarrh at the time of the examination; twelve of these men considered that it had developed in the subway. Catarrh was slightly more prevalent among the motormen than conductors.

In seventy-seven cases the catarrhal secretion was black, brown, green, or dirty. There was no difference between the conductors, motormen and switchmen in this particular. It was described as from the throat in all cases except one. In this exception it was said to be from the stomach — an obvious impossibility.

Pharyngitis occurred in seventy-two cases, of which fifty-three were acute or sub-acute. There was about as much pharyngitis among the motormen as among the conductors.

Laryngitis occurred in eight cases. A slight congestion was present in forty-seven cases.

Examination of the lungs. Cough was present in twenty-eight cases. In only one case was it considerable. The chest configuration was " good " or " excellent " in eighty cases. It was " poor " in one case. The average circumference was 90.1 centimeters; the lowest 80 centimeters.

Six per cent of the men gave a family history of tuberculosis. There had been five cases of bronchitis, three of which occurred among the conductors.

Slight emphysema, antedating the beginning of subway employment, occurred in two cases, both among conductors.

In thirteen cases slight pains were described. They were usually in the side or in the shoulder.

Dry pleurisy was present in fifty-three cases. It was axillary in forty-four cases; bilateral in seventeen cases; at the apex in twelve cases. It was distributed proportionately among the three classes of men examined.

Slight infiltration was present in thirteen cases. It was combined with pleurisy in eight cases.

There were five cases of slight fibrosis, all among the motormen. Three of these cases were described as doubtful; they were combined with pleurisy.

Of all these conditions the prevalence of dry pleurisy seemed worthy of further study.

RESULTS OF ANALYSES OF SPUTUM, URINE, AND SWEAT

Laboratory tests were made of sputum, urine, and sweat with the object of throwing light upon the findings of the medical examinations.

The specimens of sputum were, in every case, stained and examined microscopically for tubercle bacilli. None was found. The presence of other bacteria and macerated epithelium from the mouth was frequently noted, but proved nothing of importance.

Specimens of sputum were examined for iron. After some experiment with different methods, the test adopted was the digestion of fresh sputum with strong hydrochloric acid and the addition of ammonium sulphocyanide. Particles of iron were frequently found, but it was impossible to say that they came from the lungs. In fact, in practically every case the specimens of sputum were only secretions from the mouth and throat. Satisfactory specimens from the bronchi could probably only be obtained, if at all, from the first expectoration of the early morning. Such specimens were requested of many of the men, but were never furnished.

A few samples of urine were examined for iron. The test employed was the same as that used in examining sputum. No evidence of iron was discovered. If present at all, the amount was extremely slight.

The sweat was examined for iron for the reason that the underclothing of many of the employees became stained yellow where the ordinary marks of perspiration might alone be expected. The color of these stains suggested iron. Apparently the stains came from iron dust dissolved upon the surface of the body and were not due to the condition of the sweat which was exuded from the skin. It was impossible to tell from the tests the source of the iron, but the fact that the bodies of most of the men contained iron particles, and the fact that the samples of sweat were not free from skin dirt, probably offers sufficient explanation of this condition.

RESULTS OF THE AUTOPSIES

Autopsies were performed upon the bodies of five employees and one other person killed in the subway during the year 1906. Four of the employees were trackmen; one had been employed in the subway six months, two two months, and one three months. In addition, there were a switchman and a guard who had been employed a year each. All were of fine physique and below forty years of age. Most of the men were Italians and had come from Italy, where they had led an outdoor life.

The autopsies showed but few of the conditions which

have been described in medical literature as characteristic of siderosis and other dust diseases. The bodies were usually black with the peculiar dust of the subway, but a surprisingly small amount of this dust was found within. Iron particles were extremely hard to find in the trachea, bronchioles, and air cells of the lungs, notwithstanding the fact that the men had been run over and had probably gasped dusty air directly in through the open mouth while expiring. The air passages were invariably in a normal condition considering the fact that the men were city dwellers.

Under the microscope and with proper staining methods the lymphatics were seen to contain metallic iron, but not in overwhelming amount. Particles of iron could now and then be found in the alveoli. With the iron particles were masses of other comminuted foreign matter, chiefly soot. The lungs had lost nothing of their spongy character. There were no bands of hard fibrous tissue running through them, as might be expected in siderosis. The walls of the bronchial tubes were not thickened.

A slight diffuse pleurisy was found in all cases, but iron particles did not exist in greater amount in the areas affected by this pleurisy than elsewhere. It is doubtful whether any of these pleurisies could have been discovered before death.

The test for iron was a refinement of the familiar hydrochloric acid-and-potassium ferrocyanide method performed upon histological sections. The iron particles were also recovered by incineration.

The lungs of the subway employees autopsied contained somewhat more iron than a lung which was assumed to be normal, but this normal lung contained no metallic iron at all. Inasmuch as it is probable that the lungs of all persons who have lived a few months in New York contain iron particles in sufficient number to be detectable by the delicate method of analysis employed, it seems likely that the standard of comparison used was unreasonably severe. It may be said, therefore, that the autopsies threw no light either upon the possibly evil effects of the dust or the prevalence of dry pleurisy.

It would be equally unfair to assume from this analytical evidence that the dust was or was not injurious when breathed under the circumstances which surrounded these men. To settle this question would require many more autopsies, and it would be essential to have them performed upon the bodies of persons who have been longer exposed to the air. This was, of course, impossible under the circumstances.

POSSIBLE CAUSES AND CONSEQUENCES OF THE PLEURISY FOUND

The fact that a very large amount of pleurisy existed among the subway employees had not been anticipated. It could not at once be explained. Its importance depended upon whether it was due to dust or other condition peculiar to the subway, and whether it was associated with some physiological condition still more serious.

It became desirable to inquire very carefully into the nature of the pleurisy and the conditions with which it was connected. These studies were too extensive to be fully reported here, but it seems desirable that some of their more essential features should be recorded.

Pleurisy, or pleuritis, as it is more accurately called, is an inflammation, or the result of an inflammation, of the pleural membrane which surrounds the lungs. This membrane has been likened to two sacks, which are partly in contact, one within the other. The inner pleura closely covers the lungs, while the outer lines the ribs and other tissues of the chest cavity. In health the surfaces of the two pleura are very smooth and glide over one another without perceptible friction as the lungs expand and contract in breathing. In pleurisy this smoothness disappears and inflammation occurs. Eventually the opposing surfaces become rough or adherent, a condition which can be detected when a stethoscope is applied to the outside of the chest wall. When the friction sounds are very pronounced, they are technically described as friction rubs, and when less so, crepitant rales. Other terms are sometimes used to describe the sounds, but they are all practically some modification of these.

Dry pleurisy, when chronic, is, in most cases, the result of the more common acute pleurisy with effusion, yet there is a primitive dry pleurisy which may occur without any of the symptoms which generally accompany pleurisy of the latter sort. Pain and a characteristic cough usually call attention to the existence of pleurisy. It was a remarkable fact, repeatedly noted with surprise by the medical examiners, that the subway employees rarely complained of pain, cough, or any other of the clinical symptoms of pleurisy.

The causes of pleurisy are believed to be generally microbic. Several kinds of harmful bacteria have been known to reach the pleura and set up inflammation. It is also a frequent complication in pneumonia and bronchitis and is associated to some extent with tuberculosis.

The frequency with which pleurisy is noted in autopsy in connection with other diseases of the respiratory organs is shown in Table XII, made from data kindly supplied by Professor F. B. Mallory, Associate Professor of Pathology, Harvard Medical School.

TABLE XII

FREQUENCY WITH WHICH PLEURISY WAS FOUND AT AUTOPSY IN 1008 CASES OF OTHER RESPIRATORY DISEASES AT THE BOSTON CITY HOSPITAL, 1901–1905

Year.	Number of autopsies.	Number of cases of lobar pneumonia.	Number of cases of broncho-pneumonia.	Number of cases of acute pleurisy.	Number of cases of chronic pleurisy.	Number of cases of tuberculosis of lungs.	Number of cases of tuberculous pleurisy.
1901	177	32	37	34	110	44	...
1902	213	24	40	12	94	23	1
1903	211	23	38	30	113	46	1
1904	200	24	72	35	117	45	1
1905	207	27	33	29	84	19	2
Total	1,008	130	220	140	518	167	5

This table shows that pleurisy existed, with other respiratory diseases, to the extent of 65.8 per cent. In most cases these other diseases were probably the inciting cause of the pleurisy.

In a way not yet entirely explained a sudden chill is an important factor in producing pleurisy. A slight dry pleurisy may follow almost immediately upon exposure. The onset may resemble the onset of pleurisy with effusion, yet, after a few days, the symptoms disappear and no effusion occurs. A large percentage of the pleuritic adhesions seen after death are believed to originate in this way.

Dry pleurisy is never fatal. Extensive adhesions, it appears, may somewhat interfere with the normal action of the lungs, but if they do so their effects are not serious. The importance of dry pleurisy depends chiefly upon other diseases. People die of the diseases which lead to the pleurisy. In seeking to explain the condition of the subway employees, therefore, it seemed desirable to look carefully for respiratory affections.

Pleurisy among the subway employees. For purposes of study, the records of the cases of dry pleurisy among the subway employees were gathered together into two groups, according to the friction sounds by which the pleurisy had been diagnosed. These were designated: Group I, Pleuritic Crepitations, and Group II, Pleuritic Rubs.

Cases of dry pleurisy in men who reported that they had experienced an attack of pleurisy or pneumonia before entering the subway were considered sufficiently accounted for and excluded from further study. This reduced the number from fifty-three to forty-five cases. Finally, one case, suspicious of a former attack of tuberculosis, was excluded, leaving forty-four cases of dry pleurisy unaccounted for.

The following data show where and to what extent friction sounds were heard and the condition of the nose and throat in each group:

GROUP I. PLEURITIC CREPITATIONS; TWENTY-SEVEN CASES.

DISTINCTNESS OF CREPITATIONS.		LOCATION OF CREPITATIONS.	
Very distinct	1	Apex	3
Distinct	12	Axilla	20
Slight	14	Apex and axilla	4
	27		27

With these the following conditions of the nose and throat were noted:

RHINITIS.		PHARYNGITIS.		LARYNGITIS.	
Marked	2	Marked	0	Marked	1
Present	8	Present	1	Present	10
Slight	7	Slight	13	Slight	11
	17		14		22

The foregoing data show that the pleurisy was located mostly in the axilla. It was generally accompanied by pharyngitis and slight laryngitis; rhinitis was present in about half the cases in Group I.

GROUP II. PLEURITIC RUBS; SEVENTEEN CASES.

DISTINCTNESS OF RUBS.	
Very distinct	6
Distinct	11
Slight	0
	17

LOCATION OF RUBS.	
Apex	2
Right axilla	5
Left axilla	10
	17

The following conditions of the nose and throat were noted in these cases:

RHINITIS.	
Marked	2
Present	3
Slight	11
	16

PHARYNGITIS.	
Marked	2
Present	4
Slight	10
	16

LARYNGITIS.	
Marked	0
Present	1
Slight	11
	12

It will be observed that in Group II the pleurisy was located chiefly in the axilla, more often in the left than in the right side. It was in all but one case accompanied by slight rhinitis, and usually by pharyngitis and laryngitis.

There is considerable similarity between the data thus collated for the two groups. The friction sounds were found in the axilla as a rule. In more than half the cases there was congestion or inflammation of the nose and throat, this condition being very slight among the cases contained in Group II, in which the pleurisy was most marked.

A rather large amount of congestion and inflammation of the nose and throat existed among the employees who were apparently quite free from pleurisy. The following data illustrate this by showing the frequency with which laryngitis, pharyngitis, and rhinitis occurred among all the employees included in this study:

Number of men.	Employees.	Rhinitis.		Pharyngitis.		Laryngitis.	
		Cases.	Per cent.	Cases.	Per cent.	Cases.	Per cent.
47	Without pleurisy .	28	60	29	62	24	51
44	With pleurisy . .	34	77	34	77	21	47
91	Difference	6	17	5	15	3	4

These studies showed that the condition of the men with and without pleurisy was, in almost all ways, identical, although among those who had pleurisy there was a little more rhinitis and pharyngitis but less laryngitis than among those who were without it.

Among the forty-four cases only one complained of sore throat, although seven said that their throats became slightly dry, and seven complained of hoarseness. In thirty-five cases the men spoke of a slight expectoration, thirty describing it as "black," "gray," "dark," "dirty," or "green," and three "white." Pain was mentioned in eight cases. It was always described as slight or occasional. In five instances the pain was in the shoulder, or axilla, and in two cases in the chest.

Normal amount of pleurisy among city dwellers. In order to determine just how excessive was the prevalence of dry pleurisy among the subway employees, it seemed desirable to inquire how often this disease occurred among persons engaged in other occupations.

It was well to know that dry pleurisy in its milder forms frequently existed among persons in good health and that it was extremely rare to find a human body after death free from a roughened or adherent pleura, but just how commonly pleurisy occurred to the extent noted among subway employees could not be determined from the literature of the subject. It had to be sought by special investigation.

Two hundred persons were, therefore, subjected to a physical examination of the lungs similar to that given the subway employees. The work was done by the same principal examiner. The men examined represented a large number of vocations and were chosen at random from among persons admitted to Bellevue Hospital for various causes.

It was found that dry pleurisy existed in the same degree as met with among subway employees to the extent of 14½ per cent. If allowance had been made for their medical histories, deducting old pleurisies, emphysema, and pneumonia from the count, as had been done in studying the records relating to the subway employees, this percentage undoubtedly would have been slightly reduced.

Among the two hundred outsiders the pleuritic sounds were noted in the axilla in twenty-two cases and in the apex in seven cases. This was about the same ratio as found among the subway employees. The diagnostic signs noted were crepitant rales twenty-five times and friction rubs four times. Among the subway employees the occurrence of rubs was relatively more frequent; in other works, the pleurisy was more marked.

The cases of dry pleurisy found among two hundred men engaged in various employments are shown in Table XIII.

TABLE XIII

OCCUPATIONS OF PERSONS FOUND TO HAVE DRY PLEURISY AMONG TWO HUNDRED PERSONS TAKEN AT RANDOM

Occupation.	Cases of pleurisy.	Occupation.	Cases of pleurisy.
Butcher	2	Street cleaner	1
Driver	3	Conductor	1
Laborer	4	Real estate agent	1
Iron worker	2	Janitor	1
Steam fitter	1	Poleman	1
Clerk	1	Gardener	1
Porter	1	Tunnel worker	1
Brass polisher	1	Laundryman	1
Horse clipper	2	Orderly	1
Cook	1	Oysterman	1
Roofer	1		
	19		10

Of these twenty-nine cases, six were among persons engaged in dusty work.

The records relating to the forty-four cases of pleurisy which had not thus far been explained were next examined in the hope of determining whether a knowledge of the previous occupations of the employees would throw any light upon their condition. This study proved more satisfactory than was anticipated. The leading facts of interest concerning the histories of the men are given in Table XIV, which is divided into two parts in accordance with the severity of the pleurisy as indicated by the diagnostic signs, pleuritic rubs, and pleuritic crepitations.

Of the seventeen employees in whom dry pleurisy was diagnosed by pleuritic rubs, six had, previous to their subway employment, been steam locomotive engineers, firemen, or brakemen on outside railroads, with an average period of service in these vocations of 10.1 years each. Of the remaining eleven, six had been motormen or conductors with an average period of service of 8.6 years. Of the remaining five, one had been an iron worker for eighteen years.

Of the twenty-seven employees whose pleurisy was diagnosed by pleuritic crepitations, nine had been steam locomotive engineers, firemen, or brakemen, with an average term of railway service of 7.4 years. Of the remaining eighteen, seven had been motormen, conductors, or switchmen, with an average period of 10.4 years.

TABLE XIV

LENGTH OF SERVICE IN THE SUBWAY AND IN PREVIOUS OCCUPATIONS OF EMPLOYEES WITH PLEURISY

PART I. — *Diagnostic Sign — Pleuritic Rubs*

No.	Subway employment.		Previous employment.	
	Grade.	Period of service.	Occupation.	Period of service.
		Mo.		Yrs.
1	Motorman	20	Locomotive fireman	6
2	Motorman	21	Conductor	11
3	Motorman	19	Locomotive fireman (5), motorman (3)	8
4	Motorman	19	Motorman	5
5	Motorman	15	Storekeeper	19
6	Motorman	21	Locomotive engineer	14
7	Motorman	21	Locomotive fireman (7), engineer (9)	16
8	Motorman	9	Motorman	10
9	Motorman	22	Locomotive fireman (9), engineer (5)	14
10	Motorman	17	Locomotive fireman (3), brakeman (3)	6
11	Motorman	22	Motorman	5
12	Conductor	20	Janitor	8–9
13	Conductor	16	Motorman	9
14	Conductor	16	Dry goods	8
15	Conductor	16	Motorman (1½), furrier (12)	13½
16	Conductor	26	Tamper	8
17	Conductor	16	Iron worker	17–18

TABLE XIV — *Continued*

LENGTH OF SERVICE IN THE SUBWAY AND IN PREVIOUS OCCUPATIONS OF EMPLOYEES WITH PLEURISY

PART II. — *Diagnostic Sign — Pleuritic Crepitations*

	Subway employment.		Previous employment.	
No.	Grade.	Period of service.	Occupation.	Period of service.
		Mo.		Yrs.
1	Motorman	19	Motorman	3
2	Motorman	19	Conductor	10
3	Motorman	21	Locomotive fireman (3), storekeeper (6)	9
4	Motorman	15	Locomotive fireman (3), iron molder (5), steam fitter (11)	19
5	Motorman	21	Locomotive fireman (3), brakeman (4)	7
6	Motorman	9	Locomotive fireman (9), engineer (7)	16
7	Motorman	24	Motorman	20
8	Motorman	10	Locomotive fireman (18), motorman (4)	22
9	Motorman	16	Locomotive fireman (3), brakeman (3), conductor (3), yardmaster (3)	12
10	Motorman	22	Switchman	18
11	Motorman	22	Brakeman	10
12	Switchman	22	Office clerk	10
13	Switchman	16	Locomotive fireman	2
14	Switchman	15	. . . .	. . .
15	Switchman	16	Expressman	5
16	Switchman	16	Conductor (trolley)	4
17	Conductor	22	Clerk	7½
18	Conductor	19	Elevator man	5
19	Conductor	24	Inspector (2), bartender (7)	9
20	Conductor	16	Conductor (5), salesman (14). . . .	19
21	Conductor	16	Locomotive fireman (5), railroads (20)	25
22	Conductor	16	Machinist	12
23	Conductor	16	Conductor (trolley)	13
24	Conductor	16	Leather goods	7
25	Conductor	21	Clerk (1), contractor (19)	20
26	Conductor	16	Rubber goods (1½), truck gardener (5)	6½
27	Conductor	20	Fireman (2), grocery clerk (10) . .	12

Here were twenty-eight men who had, previous to their subway work, been engaged in employments in which they were exposed to alternate heat and cold to a remarkable and unusual extent and for an average period of nine years each.

Cause of the pleurisy. It seemed impossible to avoid the conclusion that the excessive amount of dry pleurisy was largely the result of forgotten or unrecognized attacks of pleurisy experienced by the men before they entered upon their subway work.

Many circumstances favored this opinion, among which may be mentioned the absence of pain or other clinical symptoms of acute pleurisy, absence of pneumonia, tuberculosis, or bronchitis, and the excellent health records of the men since going to work in the subway. Furthermore, there was as much pleurisy among the motormen as among the conductors, while their exposure both to draughts and dust was quite different: the motormen were shut up in their small compartments and were well protected, while the conductors standing between the cars were much exposed.

Subtracting the twenty-eight cases of pleurisy thus explained from the forty-four cases which had been without explanation left sixteen cases finally unaccounted for. This was about the normal.

CONCLUSIONS

The principal conclusions reached by the author concerning the various subjects dealt with in this investigation follow:

1. The air of the subway, as judged by analyses and by careful studies of the health of the men, was not injurious. If injury was being done, the subway had not been in operation long enough and the investigation had been too short to discover it.

2. The most objectionable feature of the air was the dust, which consisted chiefly of angular particles of iron. It was possible also that injurious bacteria might sometimes be associated with these metallic particles. Lack of strict enforcement of the city ordinance against spitting and the want of skillful care in cleaning the subway, made this danger greater than it need be.

3. The odor and heat of the subway, although very disagreeable, were not actually injurious to health. The strong draughts and changes of temperature which occurred at the stations were the most objectionable atmospheric conditions, so far as health was concerned.

4. The employees submitted by the company for physical examination were a particularly robust lot of men. From their excellent physique it appeared that they had been carefully selected, a fact which was explained when it was found that a large majority of the men had previously been engaged in railroading, where capacity to do hard, manual labor was required. It was fair to assume that the employees examined represented a fair average of all those who came in close contact with the passengers, so far as resistance to disease was concerned.

5. There had been very little sickness among the employees during their period of subway employment, judging by the accounts which the men gave. No information with respect to this matter was obtainable from the operating company. Many of the men claimed to have gained weight since they had been working in the subway — a fact due, apparently, less to any peculiarly healthful property of the air than to the easier work required.

6. Most of the men spoke of drowsiness. This was perhaps to be explained by the comparative darkness of the subway, the monotony of the work, and fatigue of the

eyes. The drowsiness was apparently never sufficient to keep the men from performing their duties properly.

7. A large number of employees complained of yellow stains which came upon their underclothing, as they supposed, from their sweat. This caused considerable inconvenience. The stains probably resulted from iron particles upon the body which were acted upon by the sweat. Investigation excluded the possibility that the sweat itself was discolored.

8. Careful physical examinations showed that an excessive amount of dry pleurisy, without pain or other physical discomfort, existed among the men. Pleurisy occurred to the extent of 53 per cent among the employees and to the extent of 14½ per cent among persons not engaged in subway work.

The cause of the dry pleurisy was not at first apparent, but upon investigation it appeared to have been in no way due to the subway. Nine per cent of the men had medical histories which accounted for their condition, and 28 per cent had worked for many years under conditions known to be favorable to the occurrence of this disease. The pleurisy had no visible effect upon the health of the men and was not likely to be injurious to them in the future.

9. Congestion and inflammation of the upper air passages were prevalent. Rhinitis and pharyngitis in acute or chronic form occurred in about 70 per cent of the men examined. Laryngitis was less common, occurring in about 55 per cent. These figures are somewhat above the normal, considering the degree of severity represented. No case of bronchitis was discovered. The prevalence of the minor respiratory affections noted was due, apparently, more to the previous employments of the men than to their present surroundings, although the excessive use of the

voice required of the conductors seemed likely to aggravate these affections.

10. Analyses of the sputum, urine, and sweat of the men showed that iron dust was given off only in the sputum. This sputum was derived mostly from the mouth and throat, where most of the iron particles drawn in with the inspired air were caught.

11. The findings at autopsy threw no light upon the possibly evil effects of the dust. The men whose bodies were examined had worked too short a time in the subway for information of value in this direction to be obtainable. Iron was found in the lungs of all, but to an extent which had produced no evil consequences.

RECOMMENDATIONS

Certain specific recommendations seemed to be required, under the circumstances.

1. Care should be taken that persons employed in the subway are free from respiratory disease or a tendency toward it. This rule should be extended to all grades and positions and made to apply, also, to the women who operate the news stands.

2. Thorough physical examinations, especially of the respiratory apparatus and heart, should be made of all employees when they are first engaged and at yearly intervals subsequently.

3. While the dust was not proved to have produced harmful results, sanitary considerations require that it should be prevented as far as practicable from getting into the air. To this end (*a*) sand and sawdust should not purposely be scattered on the stairways and platforms; (*b*) sweeping and cleaning should be done in a strictly sanitary manner, preferably in accordance with the recom-

mendations of the Advisory Board of the Department of Health; and (*c*) investigations should be made to determine whether it is feasible to prevent or collect much of the iron dust.

4. The city ordinance against spitting should be enforced to the letter. Although some progress has already been made in preventing it, spitting is still practiced occasionally on the platforms and on the roadbed. Not only passengers, but employees are offenders in this direction.

INDEX

LIBRARY

SHORT-TITLE CATALOGUE

OF THE

PUBLICATIONS

OF

JOHN WILEY & SONS,

NEW YORK.

LONDON: CHAPMAN & HALL, LIMITED.

ARRANGED UNDER SUBJECTS.

Descriptive circulars sent on application. Books marked with an asterisk (*) are sold at *net* prices only. All books are bound in cloth unless otherwise stated.

AGRICULTURE—HORTICULTURE—FORESTRY.

Armsby's Manual of Cattle-feeding....................12mo, $1 75
  Principles of Animal Nutrition....................8vo, 4 00
Budd and Hansen's American Horticultural Manual:
  Part I. Propagation, Culture, and Improvement....................12mo, 1 50
  Part II. Systematic Pomology....................12mo, 1 50
Elliott's Engineering for Land Drainage....................12mo, 1 50
  Practical Farm Drainage....................12mo, 1 00
Graves's Forest Mensuration....................8vo, 4 00
Green's Principles of American Forestry....................12mo, 1 50
Grotenfelt's Principles of Modern Dairy Practice. (Woll.)....................12mo, 2 00
* Herrick's Denatured or Industrial Alcohol....................8vo, 4 00
Kemp and Waugh's Landscape Gardening. (New Edition, Rewritten. In Preparation).
* McKay and Larsen's Principles and Practice of Butter-making.......8vo, 1 50
Maynard's Landscape Gardening as Applied to Home Decoration.....12mo, 1 50
Quaintance and Scott's Insects and Diseases of Fruits. (In Preparation).
Sanderson's Insects Injurious to Staple Crops....................12mo, 1 50
* Schwarz's Longleaf Pine in Virgin Forests....................12mo, 1 25
Stockbridge's Rocks and Soils....................8vo, 2 50
Winton's Microscopy of Vegetable Foods....................8vo, 7 50
Woll's Handbook for Farmers and Dairymen....................16mo, 1 50

ARCHITECTURE.

Baldwin's Steam Heating for Buildings....................12mo, 2 50
Berg's Buildings and Structures of American Railroads....................4to, 5 00
Birkmire's Architectural Iron and Steel....................8vo, 3 50
  Compound Riveted Girders as Applied in Buildings....................8vo, 2 00
  Planning and Construction of American Theatres....................8vo, 3 00
  Planning and Construction of High Office Buildings....................8vo, 3 50
  Skeleton Construction in Buildings....................8vo, 3 00
Briggs's Modern American School Buildings....................8vo, 4 00
Byrne's Inspection of Material and Wormanship Employed in Construction. 16mo, 3 00
Carpenter's Heating and Ventilating of Buildings....................8vo, 4 00

Fireproofing of Steel Buildings. 8vo, 2 50
French and Ives's Stereotomy. 8vo, 2 50
Gerhard's Guide to Sanitary House-Inspection. 16mo, 1 00
* Modern Baths and Bath Houses 8vo, 3 00
Sanitation of Public Buildings 12mo, 1 50
Theatre Fires and Panics. 12mo, 1 50
Holley and Ladd's Analysis of Mixed Paints, Color Pigments, and Varnishes
Large 12mo, 2 50
Johnson's Statics by Algebraic and Graphic Methods 8vo, 2 00
Kellaway's How to Lay Out Suburban Home Grounds 8vo, 2 00
Kidder's Architects' and Builders' Pocket-book. .... 16mo, mor. 5 00
Maire's Modern Pigments and their Vehicles 12mo, 2 00
Merrill's Non-metallic Minerals: Their Occurrence and Uses. 8vo, 4 00
Stones for Building and Decoration. 8vo, 5 00
Monckton's Stair-building. 4to, 4 00
Patton's Practical Treatise on Foundations. 8vo, 5 00
Peabody's Naval Architecture 8vo, 7 50
Rice's Concrete-block Manufacture 8vo, 2 00
Richey's Handbook for Superintendents of Construction. .... 16mo, mor. 4 00
* Building Mechanics' Ready Reference Book:
* Building Foreman's Pocket Book and Ready Reference. (In Press.)
* Carpenters' and Woodworkers' Edition. .... 16mo, mor. 1 50
* Cement Workers and Plasterer's Edition. .... 16mo, mor. 1 50
* Plumbers', Steam-Filters', and Tinners' Edition. .... 16mo, mor. 1 50
* Stone- and Brick-masons' Edition. .... 16mo, mor. 1 50
Sabin's House Painting 12mo, 1 00
Industrial and Artistic Technology of Paints and Varnish. 8vo, 3 00
Siebert and Biggin's Modern Stone-cutting and Masonry. 8vo, 1 50
Snow's Principal Species of Wood. 8vo, 3 50
Towne's Locks and Builders' Hardware. .... 18mo, mor. 3 00
Wait's Engineering and Architectural Jurisprudence 8vo, 6 00
Sheep, 6 50
Law of Contracts. 8vo, 3 00
Law of Operations Preliminary to Construction in Engineering and Architecture. 8vo, 5 00
Sheep, 5 50
Wilson's Air Conditioning 12mo, 1 50
Worcester and Atkinson's Small Hospitals, Establishment and Maintenance, Suggestions for Hospital Architecture, with Plans for a Small Hospital.
12mo, 1 25

ARMY AND NAVY.

Bernadou's Smokeless Powder, Nitro-cellulose, and the Theory of the Cellulose Molecule 12mo, 2 50
Chase's Art of Pattern Making. 12mo, 2 50
Screw Propellers and Marine Propulsion. 8vo, 3 00
* Cloke's Enlisted Specialist's Examiner 8vo, 2 00
Gunner's Examiner 8vo, 1 50
Craig's Azimuth. 4to, 3 50
Crehore and Squier's Polarizing Photo-chronograph. 8vo, 3 00
* Davis's Elements of Law. 8vo, 2 50
* Treatise on the Military Law of United States. 8vo, 7 00
Sheep, 7 50
De Brack's Cavalry Outpost Duties. (Carr). .... 24mo, mor. 2 00
* Dudley's Military Law and the Procedure of Courts-martial. .. Large 12mo, 2 50
Durand's Resistance and Propulsion of Ships. 8vo, 5 00

Eissler's Modern High Explosives. 8vo, 4 00
* Fiebeger's Text-book on Field Fortification. .Large 12mo, 2 00
Hamilton and Bond's The Gunner's Catechism 18mo, 1 00
* **Hoff's Elementary Naval Tactics**. 8vo, 1 50
Ingalls's Handbook of Problems in Direct Fire. 8vo, 4 00
* **Lissak's Ordnance and Gunnery**. 8vo, 6 00
* Ludlow's Logarithmic and Trigonometric Tables 8vo, 1 00
* Lyons's Treatise on Electromagnetic Phenomena. Vols. I. and II.. 8vo, each, 6 00
* Mahan's Permanent Fortifications. (Mercur) . . . . .8vo, half mor. 7 50
Manual for Courts-martial. . . . . .16mo, mor. 1 50
* Mercur's Attack of Fortified Places. 12mo, 2 00
* Elements of the Art of War. 8vo, 4 00
Metcalf's Cost of Manufactures—And the Administration of Workshops. .8vo, 5 00
Nixon's Adjutants' Manual. 24mo, 1 00
Peabody's Naval Architecture. 8vo, 7 50
* Phelps's Practical Marine Surveying. 8vo, 2 50
Putnam's Nautical Charts. 8vo, 2 00
Sharpe's Art of Subsisting Armies in War. . . . . .18mo, mor. 1 50
* Tupes and Poole's Manual of Bayonet Exercises and Musketry Fencing.
24mo, leather, 50
* Weaver's Military Explosives. 8vo, 3 00
Woodhull's Notes on Military Hygiene. 16mo, 1 50

ASSAYING.

Betts's Lead Refining by Electrolysis. 8vo, 4 00
Fletcher's Practical Instructions in Quantitative Assaying with the Blowpipe.
16mo, mor. 1 50
Furman's Manual of Practical Assaying. 8vo, 3 00
Lodge's Notes on Assaying and Metallurgical Laboratory Experiments. . . .8vo, 3 00
Low's Technical Methods of Ore Analysis. 8vo, 3 00
Miller's Cyanide Process. 12mo, 1 00
Manual of Assaying. 12mo, 1 00
Minet's Production of Aluminum and its Industrial Use. (Waldo). 12mo, 2 50
O'Driscoll's Notes on the Treatment of Gold Ores. 8vo, 2 00
Ricketts and Miller's Notes on Assaying. 8vo, 3 00
Robine and Lenglen's Cyanide Industry. (Le Clerc). .8vo, 4 00
Ulke's Modern Electrolytic Copper Refining. 8vo, 3 00
Wilson's Chlorination Process. 12mo, 1 50
Cyanide Processes. 12mo, 1 50

ASTRONOMY.

Comstock's Field Astronomy for Engineers. 8vo, 2 50
Craig's Azimuth 4to, 3 50
Crandall's Text-book on Geodesy and Least Squares. 8vo, 3 00
Doolittle's Treatise on Practical Astronomy. 8vo, 4 00
Gore's Elements of Geodesy. 8vo, 2 50
Hayford's Text-book of Geodetic Astronomy. 8vo, 3 00
Merriman's Elements of Precise Surveying and Geodesy. 8vo, 2 50
* Michie and Harlow's Practical Astronomy. 8vo, 3 00
Rust's Ex-meridian Altitude, Azimuth and Star-Finding Tables. 8vo, 5 00
* White's Elements of Theoretical and Descriptive Astronomy 12mo, 2 00

CHEMISTRY.

* Abderhalden's Physiological Chemistry in Thirty Lectures. (Hall and Defren) 8vo, 5 00
* Abegg's Theory of Electrolytic Dissociation. (von Ende)........... 12mo, 1 25
Alexeyeff's General Principles of Organic Syntheses. (Matthews)........ 8vo, 3 00
Allen's Tables for Iron Analysis.. 8vo, 3 00
Arnold's Compendium of Chemistry. (Mandel).............. Large 12mo, 3 50
Association of State and National Food and Dairy Departments, Hartford, Meeting, 1906.. 8vo, 3 00
Jamestown Meeting, 1907.. 8vo, 3 00
Austen's Notes for Chemical Students.................................. 12mo, 1 50
Baskerville's Chemical Elements. (In Preparation.)
Bernadou's Smokeless Powder.—Nitro-cellulose, and Theory of the Cellulose Molecule.. 12mo, 2 50
Bilts's Chemical Preparations. (Hall and Blanchard). (In Press.)
* Blanchard's Synthetic Inorganic Chemistry............................ 12mo, 1 00
* Browning's Introduction to the Rarer Elements........................ 8vo, 1 50
Brush and Penfield's Manual of Determinative Mineralogy............... 8vo, 4 00
* Claassen's Beet-sugar Manufacture. (Hall and Rolfe)................. 8vo, 3 00
Classen's Quantitative Chemical Analysis by Electrolysis. (Boltwood).. 8vo, 3 00
Cohn's Indicators and Test-papers...................................... 12mo, 2 00
Tests and Reagents.. 8vo, 3 00
* Danneel's Electrochemistry. (Merriam)............................... 12mo, 1 25
Dannerth's Methods of Textile Chemistry................................ 12mo, 2 00
Duhem's Thermodynamics and Chemistry. (Burgess)....................... 8vo, 4 00
Eakle's Mineral Tables for the Determination of Minerals by their Physical Properties.. 8vo, 1 25
Eissler's Modern High Explosives....................................... 8vo, 4 00
Effront's Enzymes and their Applications. (Prescott)................... 8vo, 3 00
Erdmann's Introduction to Chemical Preparations. (Dunlap).............. 12mo, 1 25
* Fischer's Physiology of Alimentation........................ Large 12mo, 2 00
Fletcher's Practical Instructions in Quantitative Assaying with the Blowpipe. 12mo, mor. 1 50
Fowler's Sewage Works Analyses... 12mo, 2 00
Fresenius's Manual of Qualitative Chemical Analysis. (Wells)........... 8vo, 5 00
Manual of Qualitative Chemical Analysis. Part I. Descriptive. (Wells) 8vo, 3 00
Quantitative Chemical Analysis. (Cohn) 2 vols..................... 8vo, 12 50
When Sold Separately, Vol. I, $6. Vol. II, $8.
Fuertes's Water and Public Health...................................... 12mo, 1 50
Furman's Manual of Practical Assaying.................................. 8vo, 3 00
* Getman's Exercises in Physical Chemistry............................. 12mo, 2 00
Gill's Gas and Fuel Analysis for Engineers............................. 12mo, 1 25
* Gooch and Browning's Outlines of Qualitative Chemical Analysis. Large 12mo, 1 25
Grotenfelt's Principles of Modern Dairy Practice. (Woll)............... 12mo, 2 00
Groth's Introduction to Chemical Crystallography (Marshall)............ 12mo, 1 25
Hammarsten's Text-book of Physiological Chemistry. (Mandel)........... 8vo, 4 00
Hanausek's Microscopy of Technical Products. (Winton).................. 8vo, 5 00
* Haskins and Macleod's Organic Chemistry.............................. 12mo, 2 00
Helm's Principles of Mathematical Chemistry. (Morgan).................. 12mo, 1 50
Hering's Ready Reference Tables (Conversion Factors)............ 16mo, mor. 2 50
* Herrick's Denatured or Industrial Alcohol............................ 8vo, 4 00
Hinds's Inorganic Chemistry.. 8vo, 3 00
* Laboratory Manual for Students....................................... 12mo, 1 00
* Holleman's Laboratory Manual of Organic Chemistry for Beginners. (Walker)... 12mo, 1 00
Text-book of Inorganic Chemistry. (Cooper)........................ 8vo, 2 50
Text-book of Organic Chemistry. (Walker and Mott).................. 8vo, 2 50

olley and Ladd's Analysis of Mixed Paints, Color Pigments, and Varnishes. Large 12mo, 2

Hopkins's Oil-chemists' Handbook........................8vo, 3

Iddings's Rock Minerals........................8vo, 5

Jackson's Directions for Laboratory Work in Physiological Chemistry..8vo, 1 50

Johannsen's Determination of Rock-forming Minerals in Thin Sections..8vo, 4 00

Johnson's Chemical Analysis of Special Steel. Steel-making. (Alloys and Graphite.) (In Press.)

Keep's Cast Iron........................8vo, 2 50

Ladd's Manual of Quantitative Chemical Analysis........................12mo, 1 00

Landauer's Spectrum Analysis. (Tingle)........................8vo, 3 00

* Langworthy and Austen's Occurrence of Aluminium in Vegetable Products, Animal Products, and Natural Waters........................8vo, 2 00

Lassar-Cohn's Application of Some General Reactions to Investigations in Organic Chemistry. (Tingle)........................12mo, 1 00

Leach's Inspection and Analysis of Food with Special Reference to State Control........................8vo, 7 50

Löb's Electrochemistry of Organic Compounds. (Lorenz)........................8vo, 3 00

Lodge's Notes on Assaying and Metallurgical Laboratory Experiments....8vo, 3 00

Low's Technical Method of Ore Analysis........................8vo, 3 00

Lunge's Techno-chemical Analysis. (Cohn)........................12mo, 1 00

* McKay and Larsen's Principles and Practice of Butter-making........8vo, 1 50

Maire's Modern Pigments and their Vehicles........................12mo, 2 00

Mandel's Handbook for Bio-chemical Laboratory........................12mo, 1 50

* Martin's Laboratory Guide to Qualitative Analysis with the Blowpipe .12mo, 60

Mason's Examination of Water. (Chemical and Bacteriological.) . .12mo, 1 25

. Water-supply. (Considered Principally from a Sanitary Standpoint. 8vo, 4 00

Mathewson's Chemical Theory for First Year College Students. (In Press).

Matthews's Textile Fibres. 2d Edition, Rewritten........................8vo, 4 00

* Meyer's Determination of Radicles in Carbon Compounds. (Tingle)..12mo, 1 25

Miller's Cyanide Process........................12mo, 1 00

Manual of Assaying........................12mo, 1 00

Minet's Production of Aluminum and its Industrial Use. (Waldo).... 12mo, 2 50

Mixter's Elementary Text-book of Chemistry........................12mo, 1 50

Morgan's Elements of Physical Chemistry........................12mo, 3 00

Outline of the Theory of Solutions and its Results........................12mo, 1 00

* Physical Chemistry for Electrical Engineers........................12mo, 1 50

Morse's Calculations used in Cane-sugar Factories........................16mo, mor. 1 50

* Muir's History of Chemical Theories and Laws........................8vo, 4 00

Mulliken's General Method for the Identification of Pure Organic Compounds. Vol. I........................Large 8vo, 5 00

O'Driscoll's Notes on the Treatment of Gold Ores........................8vo, 2 00

Ostwald's Conversations on Chemistry. Part One. (Ramsey)........12mo, 1 50

" " " Part Two. (Turnbull)........12mo, 2 00

* Palmer's Practical Test Book of Chemistry........................12mo, 1 00

* Pauli's Physical Chemistry in the Service of Medicine. (Fischer)......12mo, 1 25

* Penfield's Notes on Determinative Mineralogy and Record of Mineral Tests. 8vo, paper, 50

Tables of Minerals, Including the Use of Minerals and Statistics of Domestic Production........................8vo, 1 00

Pictet's Alkaloids and their Chemical Constitution. (Biddle)........8vo, 5 00

Podle's Calorific Power of Fuels........................8vo, 3 00

Prescott and Winslow's Elements of Water Bacteriology, with Special Reference to Sanitary Water Analysis........................12mo, 1 50

* Reisig's Guide to Piece-dyeing........................8vo, 25 00

Richards and Woodman's Air, Water, and Food from a Sanitary Standpoint..8vo, 2 00

Ricketts and Miller's Notes on Assaying........................8vo, 3 00

Rideal's Disinfection and the Preservation of Food........................8vo, 4 00

Sewage and the Bacterial Purification of Sewage........................8vo, 4 00

.... vo, 1 25
Robine and Lenglen's Cyanide Industry. (Le Clerc)...............8vo, 4 00
Ruddiman's Incompatibilities in Prescriptions...........................8vo, 2 00
Whys in Pharmacy...12mo, 1 00
Ruer's Elements of Metallography. (Mathewson) (In Preparation.)
Sabin's Industrial and Artistic Technology of Paints and Varnish........8vo, 3 00
Salkowski's Physiological and Pathological Chemistry. (Orndorff)......8vo, 2 50
Schimpf's Essentials of Volumetric Analysis..........................12mo, 1 25
* Qualitative Chemical Analysis...8vo, 1 25
Text-book of Volumetric Analysis...................................12mo, 2 50
Smith's Lecture Notes on Chemistry for Dental Students....................8vo, 2 50
Spencer's Handbook for Cane Sugar Manufacturers...................16mo, mor. 3 00
Handbook for Chemists of Beet-sugar Houses.................16mo, mor. 3 00
Stockbridge's Rocks and Soils...8vo, 2 50
* Tillman's Descriptive General Chemistry................................8vo, 3 00
* Elementary Lessons in Heat..8vo, 1 50
Treadwell's Qualitative Analysis. (Hall).................................8vo, 3 00
Quantitative Analysis. (Hall)...8vo, 4 00
Turneaure and Russell's Public Water-supplies...........................8vo, 5 00
Van Deventer's Physical Chemistry for Beginners. (Boltwood).......12mo, 1 50
Venable's Methods and Devices for Bacterial Treatment of Sewage......8vo, 3 00
Ward and Whipple's Freshwater Biology. (In Press.)
Ware's Beet-sugar Manufacture and Refining. Vol. I............Small 8vo, 4 00
" " " " " Vol. II...........Small 8vo, 5 00
Washington's Manual of the Chemical Analysis of Rocks..................8vo, 2 00
* Weaver's Military Explosives...8vo, 3 00
Wells's Laboratory Guide in Qualitative Chemical Analysis..............8vo, 1 50
Short Course in Inorganic Qualitative Chemical Analysis for Engineering Students...12mo, 1 50
Text-book of Chemical Arithmetic......................................12mo, 1 25
Whipple's Microscopy of Drinking-water...................................8vo, 3 50
Wilson's Chlorination Process...12mo, 1 50
Cyanide Processes...12mo, 1 50
Winton's Microscopy of Vegetable Foods...................................8vo, 7 50

CIVIL ENGINEERING.

BRIDGES AND ROOFS. HYDRAULICS. MATERIALS OF ENGINEERING. RAILWAY ENGINEERING.

Baker's Engineers' Surveying Instruments...............................12mo, 3 00
Bixby's Graphical Computing Table.................. Paper 19½ × 24¼ inches. 25
Breed and Hosmer's Principles and Practice of Surveying. 2 Volumes.
Vol. I. Elementary Surveying...8vo, 3 00
Vol. II. Higher Surveying...8vo, 2 50
* Burr's Ancient and Modern Engineering and the Isthmian Canal . . . 8vo, 3 50
Comstock's Field Astronomy for Engineers.................................8vo, 2 50
* Corthell's Allowable Pressures on Deep Foundations.................12mo, 1 25
Crandall's Text-book on Geodesy and Least Squares.....................8vo, 3 00
Davis's Elevation and Stadia Tables.......................................8vo, 1 00
Elliott's Engineering for Land Drainage..................................12mo, 1 50
Practical Farm Drainage..12mo, 1 00
*Fiebeger's Treatise on Civil Engineering...................................8vo, 5 00
Flemer's Phototopographic Methods and Instruments.....................8vo, 5 00
Folwell's Sewerage. (Designing and Maintenance.)........................8vo, 3 00
Freitag's Architectural Engineering..8vo, 3 50
French and Ives's Stereotomy...8vo, 2 50
Goodhue's Municipal Improvements...12mo, 1 50
Gore's Elements of Geodesy..8vo, 2 50
* Hauch's and Rice's Tables of Quantities for Preliminary Estimates . . 12mo, 1 25

Hayford's Text-book of Geodetic Astronomy. 8vo, 3 00
Hering's Ready Reference Tables. (Conversion Factors). . . . . . . .16mo, mor. 2 50
Howe's Retaining Walls for Earth. 12mo, 1 25
* Ives's Adjustments of the Engineer's Transit and Level. . . . . .16mo, Bds. 25
Ives and Hilts's Problems in Surveying . . . . .16mo, mor. 1 50
Johnson's (J. B.) Theory and Practice of Surveying. . . . . .Small 8vo, 4 00
Johnson's (L. J.) Statics by Algebraic and Graphic Methods. 8vo, 2 00
Kinnicutt, Winslow and Pratt's Purification of Sewage. (In Preparation.)
Laplace's Philosophical Essay on Probabilities. (Truscott and Emory)
12mo, 2 00
Mahan's Descriptive Geometry. 8vo, 1 50
Treatise on Civil Engineering. (1873.) (Wood). 8vo, 5 00
Merriman's Elements of Precise Surveying and Geodesy. 8vo, 2 50
Merriman and Brooks's Handbook for Surveyors. . . . . .16mo, mor. 2 00
Nugent's Plane Surveying. 8vo, 3 50
Ogden's Sewer Construction 8vo, 3 00
Sewer Design 12mo, 2 00
Parsons's Disposal of Municipal Refuse. 8vo, 2 00
Patton's Treatise on Civil Engineering. . . . . .8vo. half leather, 7 50
Reed's Topographical Drawing and Sketching 4to, 5 00
Rideal's Sewage and the Bacterial Purification of Sewage. 8vo, 4 00
Riemer's Shaft-sinking under Difficult Conditions. (Corning and Peele).. 8vo, 3 00
Siebert and Biggin's Modern Stone-cutting and Masonry. 8vo, 1 50
Smith's Manual of Topographical Drawing. (McMillan). 8vo, 2 50
Soper's Air and Ventilation of Subways . . . . .Large 12mo, 2 50
Tracy's Plane Surveying . . . . .16mo, mor. 3 00
* Trautwine's Civil Engineer's Pocket-book. .16mo, mor. 5 00
Venable's Garbage Crematories in America. 8vo, 2 00
Methods and Devices for Bacterial Treatment of Sewage. 8vo, 3 00
Wait's Engineering and Architectural Jurisprudence. 8vo, 6 00
Sheep, 6 50
Law of Contracts. 8vo, 3 00
Law of Operations Preliminary to Construction in Engineering and Architecture. 8vo, 5 00
Sheep, 5 50
Warren's Stereotomy—Problems in Stone-cutting. 8vo, 2 50
* Waterbury's Vest-Pocket Hand-book of Mathematics for Engineers.
$2\frac{7}{8} \times 5\frac{3}{8}$ inches, mor. 1 00
Webb's Problems in the Use and Adjustment of Engineering Instruments.
16mo, mor. 1 25
Wilson's (H. N.) Topographic Surveying 8vo, 3 50
Wilson's (W. L.) Elements of Railroad Track and Construction. 12mo, 2 00

BRIDGES AND ROOFS.

Boller's Practical Treatise on the Construction of Iron Highway Bridges. .8vo, 2 00
Burr and Falk's Design and Construction of Metallic Bridges 8vo, 5 00
Influence Lines for Bridge and Roof Computations. 8vo, 3 00
Du Bois's Mechanics of Engineering. Vol. II. . . . . .Small 4to, 10 00
Foster's Treatise on Wooden Trestle Bridges. 4to, 5 00
Fowler's Ordinary Foundations 8vo, 3 50
French and Ives's Stereotomy 8vo, 2 50
Greene's Arches in Wood, Iron, and Stone. 8vo, 2 50
Bridge Trusses. 8vo, 2 50
Roof Trusses. 8vo, 1 25
Grimm's Secondary Stresses in Bridge Trusses. .8vo, 2 50
Heller's Stresses in Structures and the Accompanying Deformations. 8vo, 3 00
Howe's Design of Simple Roof-trusses in Wood and Steel. 8vo, 2 00
Symmetrical Masonry Arches. 8vo, 2 50
Treatise on Arches. 8vo, 4 00

Johnson, Bryan, and Turneaure's Theory and Practice in the Designing of Modern Framed Structures..........................Small 4to, 10 00
Merriman and Jacoby's Text-book on Roofs and Bridges:
Part I. Stresses in Simple Trusses...............................8vo, 2 50
Part II. Graphic Statics...8vo, 2 50
Part III. Bridge Design..8vo, 2 50
Part IV. Higher Structures.....................................8vo, 2 50
Morison's Memphis Bridge....................................Oblong 4to, 10 00
Sondericker's Graphic Statics, with Applications to Trusses, Beams, and Arches. 8vo, 2 00
Waddell's De Pontibus, Pocket-book for Bridge Engineers...... 16mo, mor, 2 00
* Specifications for Steel Bridges.......................................12mo, 50
Waddell and Harrington's Bridge Engineering. (In Preparation.)
Wright's Designing of Draw-spans. Two parts in one volume..........8vo, 3 50

HYDRAULICS.

Barnes's Ice Formation..8vo, 3 00
Bazin's Experiments upon the Contraction of the Liquid Vein Issuing from an Orifice. (Trautwine)...8vo, 2 00
Bovey's Treatise on Hydraulics.....................................8vo, 5 00
Church's Diagrams of Mean Velocity of Water in Open Channels. Oblong 4to, paper, 1 50
Hydraulic Motors...8vo, 2 00
Mechanics of Engineering...8vo, 6 00
Coffin's Graphical Solution of Hydraulic Problems...........16mo, mor. 2 50
Flather's Dynamometers, and the Measurement of Power............12mo, 3 00
Folwell's Water-supply Engineering.....................................8vo, 4 00
Frizell's Water-power...8vo, 5 00
Fuertes's Water and Public Health......................................12mo, 1 50
Water-filtration Works...12mo, 2 50
Ganguillet and Kutter's General Formula for the Uniform Flow of Water in Rivers and Other Channels. (Hering and Trautwine)........8vo, 4 00
Hazen's Clean Water and How to Get It..........................Large 12mo, 1 50
Filtration of Public Water-supplies....................................8vo, 3 00
Hazlehurst's Towers and Tanks for Water-works.....................8vo, 2 50
Herschel's 115 Experiments on the Carrying Capacity of Large, Riveted, Metal Conduits..8vo, 2 00
Hoyt and Grover's River Discharge.....................................8vo, 2 00
Hubbard and Kiersted's Water-works Management and Maintenance.....8vo, 4 00
* Lyndon's Development and Electrical Distribution of Water Power....8vo, 3 00
Mason's Water-supply. (Considered Principally from a Sanitary Standpoint.) 8vo, 4 00
Merriman's Treatise on Hydraulics.......................................8vo, 5 00
* Michie's Elements of Analytical Mechanics..........................8vo, 4 00
* Molitor's Hydraulics of Rivers, Weirs and Sluices..................8vo, 2 00
Richards's Laboratory Notes on Industrial Water Analysis. (In Press).
Schuyler's Reservoirs for Irrigation, Water-power, and Domestic Water-supply...Large 8vo, 5 00
* Thomas and Watt's Improvement of Rivers..............................4to, 6 00
Turneaure and Russell's Public Water-supplies..........................8vo, 5 00
Wegmann's Design and Construction of Dams. 5th Ed., enlarged......4to, 6 00
Water-supply of the City of New York from 1658 to 1895..........4to, 10 00
Whipple's Value of Pure Water.................................Large 12mo, 1 00
Williams and Hazen's Hydraulic Tables.................................8vo, 1 50
Wilson's Irrigation Engineering....................................Small 8vo, 4 00
Wolff's Windmill as a Prime Mover......................................8vo, 3 00
Wood's Elements of Analytical Mechanics..............................8vo, 3 00

Baker's Roads and Pavements. 8vo, 5 00
" Treatise on Masonry Construction. 8vo, 5 00
Birkmire's Architectural Iron and Steel. 8vo, 3 50
Compound Riveted Girders as Applied in Buildings. 8vo, 2 00
Black's United States Public Works . . . . Oblong 4to. 5 00
Bleininger's Manufacture of Hydraulic Cement. (In Preparation.)
* Bovey's Strength of Materials and Theory of Structures. 8vo, 7 50
Burr's Elasticity and Resistance of the Materials of Engineering. 8vo, 7 50
Byrne's Highway Construction. 8vo, 5 00
Inspection of the Materials and Workmanship Employed in Construction. 16mo, 3 00
Church's Mechanics of Engineering. 8vo, 6 00
Du Bois's Mechanics of Engineering.
Vol. I. Kinematics, Statics, Kinetics. . . . . Small 4to, 7 50
Vol. II. The Stresses in Framed Structures, Strength of Materials and Theory of Flexures. . . . . Small 4to, 10 00
*Eckel's Cements, Limes, and Plasters. 8vo, 6 00
Stone and Clay Products used in Engineering. (In Preparation.)
Fowler's Ordinary Foundations. 8vo, 3 50
Graves's Forest Mensuration. 8vo, 4 00
Green's Principles of American Forestry. 12mo, 1 50
* Greene's Structural Mechanics. 8vo, 2 50
Holly and Ladd's Analysis of Mixed Paints, Color Pigments and Varnishes Large 12mo, 2 50
Johnson's (C. M.) Chemical Analysis of Special Steels. (In Preparation.)
Johnson's (J. B.) Materials of Construction . . . . Large 8vo, 6 00
Keep's Cast Iron. 8vo, 2 50
Kidder's Architects and Builders' Pocket-book. 16mo, 5 00
Lanza's Applied Mechanics. 8vo, 7 50
Maire's Modern Pigments and their Vehicles 12mo, 2 00
Martens's Handbook on Testing Materials. (Henning) 2 vols. 8vo, 7 50
Maurer's Technical Mechanics. 8vo, 4 00
Merrill's Stones for Building and Decoration. 8vo, 5 00
Merriman's Mechanics of Materials. 8vo, 5 00
* Strength of Materials 12mo, 1 00
Metcalf's Steel. A Manual for Steel-users. 12mo, 2 00
Morrison's Highway Engineering 8vo, 2 50
Patton's Practical Treatise on Foundations. 8vo, 5 00
Rice's Concrete Block Manufacture 8vo, 2 00
Richardson's Modern Asphalt Pavements 8vo, 3 00
Richey's Handbook for Superintendents of Construction. . . . . 16mo, mor. 4 00
* Ries's Clays: Their Occurrence, Properties, and Uses. 8vo, 5 00
Sabin's Industrial and Artistic Technology of Paints and Varnish. 8vo, 3 00
* Schwarz's Longleaf Pine in Virgin Forest 12mo, 1 25
Snow's Principal Species of Wood. 8vo, 3 50
Spalding's Hydraulic Cement. 12mo, 2 00
Text-book on Roads and Pavements. 12mo, 2 00
Taylor and Thompson's Treatise on Concrete, Plain and Reinforced. 8vo, 5 00
Thurston's Materials of Engineering. In Three Parts. 8vo, 8 00
Part I. Non-metallic Materials of Engineering and Metallurgy. 8vo, 2 00
Part II. Iron and Steel. 8vo, 3 50
Part III. A Treatise on Brasses, Bronzes, and Other Alloys and their Constituents. 8vo, 2 50
Tillson's Street Pavements and Paving Materials. 8vo, 4 00
Turneaure and Maurer's Principles of Reinforced Concrete Construction. . . 8vo, 3 00
Waterbury's Cement Laboratory Manual. 12mo, 1 00

RAILWAY ENGINEERING.

Andrews's Handbook for Street Railway Engineers........3x5 inches, mor. 1 25
Berg's Buildings and Structures of American Railroads...............4to, 5 00
Brooks's Handbook of Street Railroad Location................16mo, mor. 1 50
Butt's Civil Engineer's Field-book............................16mo, mor. 2 50
Crandall's Railway and Other Earthwork Tables...................8vo, 1 50
 Transition Curve..16mo, mor. 1 50
* Crockett's Methods for Earthwork Computations.....................8vo, 1 50
Dawson's "Engineering" and Electric Traction Pocket-book......16mo, mor. 5 00
Dredge's History of the Pennsylvania Railroad: (1879)..............Paper, 5 00
Fisher's Table of Cubic Yards...............................Cardboard, 25
Godwin's Railroad Engineers' Field-book and Explorers' Guide... 16mo, mor. 2 50
Hudson's Tables for Calculating the Cubic Contents of Excavations and Embankments..8vo, 1 00
Ives and Hilts's Problems in Surveying, Railroad Surveying and Geodesy 16mo, mor. 1 50
Molitor and Beard's Manual for Resident Engineers................16mo, 1 00
Nagle's Field Manual for Railroad Engineers..................16mo, mor. 3 00
Philbrick's Field Manual for Engineers........................16mo, mor. 3 00
Raymond's Railroad Engineering. 3 volumes.
 Vol. I. Railroad Field Geometry. (In Preparation.)
 Vol. II. Elements of Railroad Engineering......................8vo, 3 50
 Vol III. Railroad Engineer's Field Book. (In Preparation.)
Searles's Field Engineering...................................16mo, mor. 3 00
 Railroad Spiral...16mo, mor. 1 50
Taylor's Prismoidal Formulæ and Earthwork.........................8vo, 1 50
* Trautwine's Field Practice of Laying Out Circular Curves for Railroads. 12mo. mor, 2 50
* Method of Calculating the Cubic Contents of Excavations and Embankments by the Aid of Diagrams...............................8vo, 2 00
Webb's Economics of Railroad Construction..................Large 12mo, 2 50
 Railroad Construction.....................................16mo, mor. 5 00
Wellington's Economic Theory of the Location of Railways.......Small 8vo, 5 00

DRAWING.

Barr's Kinematics of Machinery....................................8vo, 2 50
* Bartlett's Mechanical Drawing...................................8vo, 3 00
* " " " Abridged Ed........................8vo, 1 50
Coolidge's Manual of Drawing...............................8vo, paper, 1 00
Coolidge and Freeman's Elements of General Drafting for Mechanical Engineers...Oblong 4to, 2 50
Durley's Kinematics of Machines...................................8vo, 4 00
Emch's Introduction to Projective Geometry and its Applications........8vo, 2 50
Hill's Text-book on Shades and Shadows, and Perspective..............8vo, 2 00
Jamison's Advanced Mechanical Drawing.............................8vo, 2 00
 Elements of Mechanical Drawing..................................8vo, 2 50
Jones's Machine Design:
 Part I. Kinematics of Machinery..................................8vo, 1 50
 Part II. Form, Strength, and Proportions of Parts...................8vo, 3 00
MacCord's Elements of Descriptive Geometry..........................8vo, 3 00
 Kinematics; or, Practical Mechanism...............................8vo, 5 00
 Mechanical Drawing..4to, 4 00
 Velocity Diagrams...8vo, 1 50
McLeod's Descriptive Geometry...........................Large 12mo, 1 50
* Mahan's Descriptive Geometry and Stone-cutting.......................8vo, 1 50
 Industrial Drawing. (Thompson.)...................................8vo, 3 50

...ing. 8vo, 1 50
Industrial Drawing. (Thompson). 8vo, 3 50
Moyer's Descriptive Geometry. 8vo, 2 00
Reed's Topographical Drawing and Sketching. 4to, 5 00
Reid's Course in Mechanical Drawing. 8vo, 2 00
Text-book of Mechanical Drawing and Elementary Machine Design. 8vo, 3 00
Robinson's Principles of Mechanism. 8vo, 3 00
Schwamb and Merrill's Elements of Mechanism. 8vo, 3 00
Smith's (R. S.) Manual of Topographical Drawing. (McMillan). 8vo, 2 50
Smith (A. W.) and Marx's Machine Design. 8vo, 3 00
* Titsworth's Elements of Mechanical Drawing . . . . Oblong 8vo, 1 25
Warren's Drafting Instruments and Operations. 12mo, 1 25
Elements of Descriptive Geometry, Shadows, and Perspective. 8vo, 3 50
Elements of Machine Construction and Drawing. 8vo, 7 50
Elements of Plane and Solid Free-hand Geometrical Drawing 12mo, 1 00
General Problems of Shades and Shadows 8vo, 3 00
Manual of Elementary Problems in the Linear Perspective of Form and Shadow 12mo, 1 00
Manual of Elementary Projection Drawing. 12mo, 1 50
Plane Problems in Elementary Geometry 12mo, 1 25
Problems, Theorems, and Examples in Descriptive Geometry. 8vo, 2 50
Weisbach's Kinematics and Power of Transmission. (Hermann and Klein). 8vo, 5 00
Wilson's (H. M.) Topographic Surveying. 8vo, 3 50
Wilson's (V. T.) Free-hand Lettering. 8vo, 1 00
Free-hand Perspective. 8vo, 2 50
Woolf's Elementary Course in Descriptive Geometry. . . . . Large 8vo, 3 00

ELECTRICITY AND PHYSICS.

* Abegg's Theory of Electrolytic Dissociation. (von Ende). 12mo, 1 25
Andrews's Hand-Book for Street Railway Engineering. . . . 3×5 inches, mor. 1 25
Anthony and Brackett's Text-book of Physics. (Magie). . . . . Large 12mo, 3 00
Anthony's Theory of Electrical Measurements. (Ball). 12mo, 1 00
Benjamin's History of Electricity. 8vo, 3 00
Voltaic Cell. 8vo, 3 00
Betts's Lead Refining and Electrolysis. 8vo, 4 00
Classen's Quantitative Chemical Analysis by Electrolysis. (Boltwood). . 8vo, 3 00
* Collins's Manual of Wireless Telegraphy. 12mo, 1 50
Mor. 2 00
Crehore and Squier's Polarizing Photo-chronograph. 8vo, 3 00
* Danneel's Electrochemistry. (Merriam). 12mo, 1 25
Dawson's "Engineering" and Electric Traction Pocket-book . . . . 16mo, mor. 5 00
Dolezalek's Theory of the Lead Accumulator (Storage Battery). (von Ende) 12mo, 2 50
Duhem's Thermodynamics and Chemistry. (Burgess). 8vo, 4 00
Flather's Dynamometers, and the Measurement of Power. 12mo, 3 00
Gilbert's De Magnete. (Mottelay). 8vo, 2 50
* Hanchett's Alternating Currents 12mo, 1 00
Hering's Ready Reference Tables (Conversion Factors) . . . . 16mo, mor. 2 50
* Hobart and Ellis's High-speed Dynamo Electric Machinery 8vo, 6 00
Holman's Precision of Measurements. 8vo, 2 00
Telescopic Mirror-scale Method, Adjustments, and Tests. . . . Large 8vo, 75
* Karapetoff's Experimental Electrical Engineering. 8vo, 6 00
Kinzbrunner's Testing of Continuous-current Machines. 8vo, 2 00
Landauer's Spectrum Analysis. (Tingle). 8vo, 3 00
Le Chatelier's High-temperature Measurements. (Boudouard—Burgess). . 12mo, 3 00
Löb's Electrochemistry of Organic Compounds. (Lorenz). 8vo, 3 00
* Lyndon's Development and Electrical Distribution of Water Power 8vo, 3 00

* Lyons's Treatise on Electromagnetic Phenomena. Vols. I. and II. 8vo, each 6 00

* Michie's Elements of Wave Motion Relating to Sound and Light...... 8vo, 4 00

Morgan's Outline of the Theory of Solution and its Results............12mo, 1 00

* Physical Chemistry for Electrical Engineers.....................12mo, 1 50

Niaudet's Elementary Treatise on Electric Batteries. (Fishback)..... 12mo, 2 50

* Norris's Introduction to the Study of Electrical Engineering.......... 8vo, 2 50

* Parshall and Hobart's Electric Machine Design.............4to, half mor. 12 50

Reagan's Locomotives: Simple, Compound, and Electric. New Edition. Large 12mo, 3 50

* Rosenberg's Electrical Engineering. (Haldane Gee — Kinzbrunner) . . 8vo, 2 00

Ryan, Norris, and Hoxie's Electrical Machinery. Vol. I............... 8vo, 2 50

Schapper's Laboratory Guide for Students in Physical Chemistry.......12mo, 1 00

* Tillman's Elementary Lessons in Heat........................... 8vo, 1 50

Tory and Pitcher's Manual of Laboratory Physics.............Large 12mo, 2 00

Ulke's Modern Electrolytic Copper Refining............................8vo, 3 00

LAW.

Brennan's Handbook: A Compendium of Useful Legal Information for Business Men...16mo, mor. 5 00

* Davis's Elements of Law..8vo, 2 50

* Treatise on the Military Law of United States.................... 8vo, 7 00

* Sheep, 7 50

* Dudley's Military Law and the Procedure of Courts-martial...Large 12mo, 2 50

Manual for Courts-martial..16mo, mor. 1 50

Wait's Engineering and Architectural Jurisprudence..................... 8vo, 6 00

Sheep, 6 50

Law of Contracts... 8vo, 3 00

Law of Operations Preliminary to Construction in Engineering and Architecture.. 8vo, 5 00

Sheep, 5 50

MATHEMATICS.

Baker's Elliptic Functions...8vo, 1 50

Briggs's Elements of Plane Analytic Geometry. (Bôcher) 12mo, 1 00

* Buchanan's Plane and Spherical Trigonometry..................... 8vo, 1 00

Byerley's Harmonic Functions.. 8vo, 1 00

Chandler's Elements of the Infinitesimal Calculus 12mo, 2 00

Coffin's Vector Analysis. (In Press.)

Compton's Manual of Logarithmic Computations......................12mo, 1 50

* Dickson's College Algebra...............................Large 12mo, 1 50

* Introduction to the Theory of Algebraic Equations.........Large 12mo, 1 25

Emch's Introduction to Projective Geometry and its Applications....... 8vo, 2 50

Fiske's Functions of a Complex Variable...............................8vo, 1 00

Halsted's Elementary Synthetic Geometry.......................... 8vo, 1 50

Elements of Geometry... 8vo, 1 75

* Rational Geometry...12mo, 1 50

Hyde's Grassmann's Space Analysis...................................... 8vo, 1 00

* Johnson's (J. B.) Three-place Logarithmic Tables: Vest-pocket size, paper, 15

100 copies, 5 00

* Mounted on heavy cardboard, 8 × 10 inches, 25

10 copies, 2 00

Johnson's (W. W.) Abridged Editions of Differential and Integral Calculus Large 12mo, 1 vol. 2 50

Curve Tracing in Cartesian Co-ordinates.........................12mo, 1 00

Differential Equations... 8vo, 1 00

Elementary Treatise on Differential Calculus............Large 12mo, 1 50

Elementary Treatise on the Integral Calculus...........Large 12mo, 1 50

Theoretical Mechanics...12mo, 3 00

Theory of Errors and the Method of Least Squares............12mo, 1 50

Johnson's Treatise on the Integral Calculus....................Large 12mo, 3 00
Treatise on Ordinary and Partial Differential Equations..Large 12mo, 3 50
Karapetoff's Engineering Applications of Higher Mathematics. (In Preparation.)
Laplace's Philosophical Essay on Probabilities. (Truscott and Emory)..12mo, 2 00
* Ludlow and Bass's Elements of Trigonometry and Logarithmic and Other Tables..........8vo, 3 00
Trigonometry and Tables published separately..........Each, 2 00
* Ludlow's Logarithmic and Trigonometric Tables..........8vo, 1 00
Macfarlane's Vector Analysis and Quaternions..........8vo, 1 00
McMahon's Hyperbolic Functions..........8vo, 1 00
Manning's Irrational Numbers and their Representation by Sequences and Series..........12mo, 1 25
Mathematical Monographs. Edited by Mansfield Merriman and Robert S. Woodward..........Octavo, each 1 00

No. 1. History of Modern Mathematics, by David Eugene Smith. No. 2. Synthetic Projective Geometry, by George Bruce Halsted. No. 3. Determinants, by Laenas Gifford Weld. No. 4. Hyperbolic Functions, by James McMahon. No. 5. Harmonic Functions, by William E. Byerly. No. 6. Grassmann's Space Analysis, by Edward W. Hyde. No. 7. Probability and Theory of Errors, by Robert S. Woodward. No. 8. Vector Analysis and Quaternions, by Alexander Macfarlane. No. 9. Differential Equations, by William Woolsey Johnson. No. 10. The Solution of Equations, by Mansfield Merriman. No. 11. Functions of a Complex Variable, by Thomas S. Fiske.

Maurer's Technical Mechanics..........8vo, 4 00
Merriman's Method of Least Squares..........8vo, 2 00
Solution of Equations..........8vo, 1 00
Rice and Johnson's Differential and Integral Calculus. 2 vols. in one. Large 12mo, 1 50
Elementary Treatise on the Differential Calculus..........Large 12mo, 3 00
Smith's History of Modern Mathematics..........8vo, 1 00
* Veblen and Lennes's Introduction to the Real Infinitesimal Analysis of One Variable..........8vo, 2 00
* Waterbury's Vest Pocket Hand-Book of Mathematics for Engineers. $2\frac{7}{8}\times5\frac{3}{8}$ inches, mor. 1 00
Weld's Determinations..........8vo, 1 00
Wood's Elements of Co-ordinate Geometry..........8vo, 2 00
Woodward's Probability and Theory of Errors..........8vo, 1 00

MECHANICAL ENGINEERING.

MATERIALS OF ENGINEERING, STEAM-ENGINES AND BOILERS.

Bacon's Forge Practice..........12mo, 1 50
Baldwin's Steam Heating for Buildings..........12mo, 2 50
Barr's Kinematics of Machinery..........8vo, 2 50
* Bartlett's Mechanical Drawing..........8vo, 3 00
* " " " Abridged Ed..........8vo, 1 50
Benjamin's Wrinkles and Recipes..........12mo, 2 00
* Burr's Ancient and Modern Engineering and the Isthmian Canal..........8vo, 3 50
Carpenter's Experimental Engineering..........8vo, 6 00
Heating and Ventilating Buildings..........8vo, 4 00
Clerk's Gas and Oil Engine..........Large 12mo, 4 00
Compton's First Lessons in Metal Working..........12mo, 1 50
Compton and De Groodt's Speed Lathe..........12mo, 1 50
Coolidge's Manual of Drawing..........8vo, paper, 1 00
Coolidge and Freeman's Elements of General Drafting for Mechanical Engineers..........Oblong 4to, 2 50

Cromwell's Treatise on Belts and Pulleys..........................12mo, 1 50
Treatise on Toothed Gearing..........................12mo, 1 50
Durley's Kinematics of Machines..........................8vo, 4 00
Flather's Dynamometers and the Measurement of Power..........................12mo, 3 00
Rope Driving..........................12mo, 2 00
Gill's Gas and Fuel Analysis for Engineers..........................12mo, 1 25
Goss s Locomotive Sparks..........................8vo, 2 00
Greene's Pumping Machinery. (In Preparation.)
Hering's Ready Reference Tables (Conversion Factors)..........................16mo, mor. 2 50
* Hobart and Ellis's High Speed Dynamo Electric Machinery..........................8vo, 6 00
Hutton's Gas Engine..........................8vo, 5 00
Jamison's Advanced Mechanical Drawing..........................8vo, 2 00
Elements of Mechanical Drawing..........................8vo, 2 50
Jones's Gas Engine. (In Press.)
Machine Design:
Part I. Kinematics of Machinery..........................8vo, 1 50
Part II. Form, Strength, and Proportions of Parts..........................8vo, 3 00
Kent's Mechanical Engineers' Pocket-book..........................16mo, mor. 5 00
Kerr's Power and Power Transmission..........................8vo, 2 00
Leonard's Machine Shop Tools and Methods..........................8vo, 4 00
* Lorenz's Modern Refrigerating Machinery. (Pope, Haven, and Dean)...8vo, 4 00
MacCord's Kinematics; or, Practical Mechanism..........................8vo, 5 00
Mechanical Drawing..........................4to, 4 00
Velocity Diagrams..........................8vo, 1 50
MacFarland's Standard Reduction Factors for Gases..........................8vo, 1 50
Mahan's Industrial Drawing. (Thompson)..........................8vo, 3 50
Oberg's Screw Thread Systems, Taps, Dies, Cutters, and Reamers. (In Press.)
* Parshall and Hobart's Electric Machine Design... Small 4to, half leather, 12 50
Peele's Compressed Air Plant for Mines..........................8vo, 3 00
Poole's Calorific Power of Fuels..........................8vo, 3 00
* Porter's Engineering Reminiscences, 1855 to 1882..........................8vo, 3 00
Reid's Course in Mechanical Drawing..........................8vo, 2 00
Text-book of Mechanical Drawing and Elementary Machine Design. 8vo, 3 00
Richard's Compressed Air..........................12mo, 1 50
Robinson's Principles of Mechanism..........................8vo, 3 00
Schwamb and Merrill's Elements of Mechanism..........................8vo, 3 00
Smith's (O.) Press-working of Metals..........................8vo, 3 00
Smith (A. W.) and Marx's Machine Design..........................8vo, 3 00
Sorel's Carbureting and Combustion in Alcohol Engines. (Woodward and Preston). Large 12mo, 3 00
Thurston's Animal as a Machine and Prime Motor, and the Laws of Energetics. 12mo, 1 00
Treatise on Friction and Lost Work in Machinery and Mill Work... 8vo, 3 00
Tillson's Complete Automobile Instructor..........................16mo, 1 50
mor. 2 00
Titsworth's Elements of Mechanical Drawing..........................Oblong 8vo, 1 25
Warren's Elements of Machine Construction and Drawing..........................8vo, 7 50
* Waterbury's Vest Pocket Hand Book of Mathematics for Engineers.
$2\frac{7}{8} \times 5\frac{3}{8}$ inches, mor. 1 00
Weisbach's Kinematics and the Power of Transmission. (Herrmann—Klein)..........................8vo, 5 00
Machinery of Transmission and Governors. (Herrmann—Klein)...8vo, 5 00
Wood's Turbines..........................8vo, 2 50

MATERIALS OF ENGINEERING

* Bovey's Strength of Materials and Theory of Structures..........................8vo, 7 50
Burr's Elasticity and Resistance of the Materials of Engineering..........................8vo, 7 50
Church's Mechanics of Engineering..........................8vo, 6 00
* Greene's Structural Mechanics..........................8vo, 2 50

Holley and Ladd's Analysis of Mixed Paints, Color Pigments, and Varnishes. Large 12mo, 2 50
Johnson's Materials of Construction..8vo, 6 00
Keep's Cast Iron. 8vo, 2 50
Lanza's Applied Mechanics...8vo, 7 50
Maire's Modern Pigments and their Vehicles.........................12mo, 2 00
Martens's Handbook on Testing Materials. (Henning)...................8vo, 7 50
Maurer's Technical Mechanics...8vo, 4 00
Merriman's Mechanics of Materials..8vo, 5 00
* Strength of Materials 12mo, 1 00
Metcalf's Steel. A Manual for Steel-users.. ....12mo, 2 00
Sabin's Industrial and Artistic Technology of Paints and Varnish........8vo, 3 00
Smith's Materials of Machines...12mo, 1 00
Thurston's Materials of Engineering............................3 vols., 8vo, 8 00
Part I. Non-metallic Materials of Engineering and Metallurgy...8vo, 2 00
Part II. Iron and Steel. 8vo, 3 50
Part III. A Treatise on Brasses, Bronzes, and Other Alloys and their Constituents..8vo, 2 50
Wood's (De V.) Elements of Analytical Mechanics.........................8vo, 3 00
Treatise on the Resistance of Materials and an Appendix on the Preservation of Timber 8vo, 2 00
Wood's (M. P.) Rustless Coatings: Corrosion and Electrolysis of Iron and Steel...8vo, 4 00

STEAM-ENGINES AND BOILERS.

Berry's Temperature-entropy Diagram.....................................12mo, 1 25
Carnot's Reflections on the Motive Power of Heat. (Thurston)........12mo, 1 50
Chase's Art of Pattern Making. 12mo, 2
Creighton's Steam-engine and other Heat-motors. 8vo, 5
Dawson's "Engineering" and Electric Traction Pocket-book.....16mo, mor. 5
Ford's Boiler Making for Boiler Makers.. ....18mo, 1
*Gebhardt's Steam Power Plant Engineering 8vo, 6
Goss's Locomotive Performance 8vo, 5
Hemenway's Indicator Practice and Steam-engine Economy..........12mo, 2 50
Hutton's Heat and Heat-engines..8vo, 5 00
Mechanical Engineering of Power Plants...................................8vo, 5 00
Kent's Steam boiler Economy..8vo, 4 00
Kneass's Practice and Theory of the Injector...............................8vo, 1 50
MacCord's Slide-valves..8vo, 2 00
Meyer's Modern Locomotive Construction....................................4to, 10 00
Moyer's Steam Turbines. (In Press.)
Peabody's Manual of the Steam-engine Indicator.........................12mo, 1 50
Tables of the Properties of Saturated Steam and Other Vapors 8vo, 1 00
Thermodynamics of the Steam-engine and Other Heat-engines......8vo, 5 00
Valve-gears for Steam-engines...8vo, 2 50
Peabody and Miller's Steam-boilers..8vo, 4 00
Pray's Twenty Years with the Indicator..............................Large 8vo, 2 50
Pupin's Thermodynamics of Reversible Cycles in Gases and Saturated Vapors. (Osterberg)...12mo, 1 25
Reagan's Locomotives. Simple, Compound, and Electric. New Edition. Large 12mo, 3 50
Sinclair's Locomotive Engine Running and Management..............12mo, 2 00
Smart's Handbook of Engineering Laboratory Practice..................12mo, 2 50
Snow's Steam-boiler Practice..8vo, 3 00
Spangler's Notes on Thermodynamics...12mo, 1 00
Valve-gears...8vo, 2 50
Spangler, Greene, and Marshall's Elements of Steam-engineering.......8vo, 3 00
Thomas's Steam-turbines..8vo

Thurston's Handbook of Engine and Boiler Trials, and the Use of the Indicator and the Prony Brake. 8vo, 5 00
Handy Tables. 8vo, 1 50
Manual of Steam-boilers, their Designs, Construction, and Operation..8vo, 5 00
Thurston's Manual of the Steam-engine. .2 vols., 8vo, 10 00
Part I. History, Structure, and Theory. 8vo, 6 00
Part II. Design, Construction, and Operation. 8vo, 6 00
Steam-boiler Explosions in Theory and in Practice. 12mo, 1 50
Wehrenfenning's Analysis and Softening of Boiler Feed-water (Patterson) 8vo, 4 00
Weisbach's Heat, Steam, and Steam-engines. (Du Bois). 8vo, 5 00
Whitham's Steam-engine Design. 8vo, 5 00
Wood's Thermodynamics, Heat Motors, and Refrigerating Machines. . .8vo, 4 00

MECHANICS PURE AND APPLIED.

Church's Mechanics of Engineering. 8vo, 6 00
Notes and Examples in Mechanics. 8vo, 2 00
Dana's Text-book of Elementary Mechanics for Colleges and Schools. .12mo, 1 50
Du Bois's Elementary Principles of Mechanics:
Vol. I. Kinematics. 8vo, 3 50
Vol. II. Statics. 8vo, 4 00
Mechanics of Engineering. Vol. I. . . . . .Small 4to, 7 50
Vol. II. . . . . .Small 4to, 10 00
* Greene's Structural Mechanics. 8vo, 2 50
James's Kinematics of a Point and the Rational Mechanics of a Particle. Large 12mo, 2 00
* Johnson's (W. W.) Theoretical Mechanics. 12mo. 3 00
Lanza's Applied Mechanics. 8vo, 7 50
* Martin's Text Book on Mechanics, Vol. I, Statics. 12mo, 1 25
* Vol. 2, Kinematics and Kinetics . .12mo, 1 50
Maurer's Technical Mechanics. 8vo, 4 00
* Merriman's Elements of Mechanics. 12mo, 1 00
Mechanics of Materials. 8vo, 5 00
* Michie's Elements of Analytical Mechanics. 8vo, 4 00
Robinson's Principles of Mechanism. 8vo, 3 00
Sanborn's Mechanics Problems. . . . . .Large 12mo, 1 50
Schwamb and Merrill's Elements of Mechanism. 8vo, 3 00
Wood's Elements of Analytical Mechanics. 8vo, 3 00
Principles of Elementary Mechanics. 12mo, 1 25

MEDICAL.

* Abderhalden's Physiological Chemistry in Thirty Lectures. (Hall and Defren) 8vo, 5 00
von Behring's Suppression of Tuberculosis. (Bolduan). 12mo, 1 00
* Bolduan's Immune Sera 12mo, 1 50
Bordet's Contribution to Immunity. (Gay). (In Preparation.)
Davenport's Statistical Methods with Special Reference to Biological Variations. . . . . .16mo, mor. 1 50
Ehrlich's Collected Studies on Immunity. (Bolduan). 8vo, 6 00
* Fischer's Physiology of Alimentation. . . . . .Large 12mo, cloth, 2 00
de Fursac's Manual of Psychiatry. (Rosanoff and Collins). . . . . .Large 12mo, 2 50
Hammarsten's Text-book on Physiological Chemistry. (Mandel). 8vo, 4 00
Jackson's Directions for Laboratory Work in Physiological Chemistry. . .8vo, 1 25
Lassar-Cohn's Practical Urinary Analysis. (Lorenz). 12mo, 1 00
Mandel's Hand Book for the Bio-Chemical Laboratory. 12mo, 1 50
* Pauli's Physical Chemistry in the Service of Medicine. (Fischer). 12mo, 1 25
* Pozzi-Escot's Toxins and Venoms and their Antibodies. (Cohn). 12mo, 1 00
Rostoski's Serum Diagnosis. (Bolduan). 12mo, 1 00
Ruddiman's Incompatibilities in Prescriptions. 8vo, 2 00

Salkowski's Physiological and Pathological Chemistry. (Orndorff)......8vo, 2 50
* Satterlee's Outlines of Human Embryology.........................12mo. 1 25
Smith's Lecture Notes on Chemistry for Dental Students.............8vo, 2 50
Steel's Treatise on the Diseases of the Dog..........................8vo, 3 50
* Whipple's Typhoid Fever.................................Large 12mo, 3 00
Woodhull's Notes on Military Hygiene.................................16mo, 1 50
* Personal Hygiene..12mo, 1 00
Worcester and Atkinson's Small Hospitals Establishment and Maintenance, and Suggestions for Hospital Architecture, with Plans for a Small Hospital..12mo, 1 25

METALLURGY.

Betts's Lead Refining by Electrolysis...................................8vo, 4 00
Bolland's Encyclopedia of Founding and Dictionary of Foundry Terms Used in the Practice of Moulding..12mo, 3 00
Iron Founder...12mo, 2 50
" " Supplement...12mo, 2 50
Douglas's Untechnical Addresses on Technical Subjects..................12mo, 1 00
Goesel's Minerals and Metals: A Reference Book.....................16mo, mor. 3 00
* Iles's Lead-smelting..12mo, 2 50
Keep's Cast Iron...8vo, 2 50
Le Chatelier's High-temperature Measurements. (Boudouard—Burgess) 12mo, 3 00
Metcalf's Steel. A Manual for Steel-users..............................12mo, 2 00
Miller's Cyanide Process...12mo, 1 00
Minet's Production of Aluminium and its Industrial Use. (Waldo)...12mo, 2 50
Robine and Lenglen's Cyanide Industry. (Le Clerc)......................8vo, 4 00
Ruer's Elements of Metallography. (Mathewson) (In Press.)
Smith's Materials of Machines..12mo, 1 00
Tate and Stone's Foundry Practice. (In Press.)
Thurston's Materials of Engineering. In Three Parts.....................8vo, 8 00
Part I. Non-metallic Materials of Engineering and Metallurgy...8vo, 2 00
Part II. Iron and Steel...8vo, 3 50
Part III. A Treatise on Brasses, Bronzes, and Other Alloys and their Constituents...8vo, 2 50
Ulke's Modern Electrolytic Copper Refining...............................8vo, 3 00
West's American Foundry Practice.......................................12mo, 2 50
Moulder's Text Book..12mo, 2 50
Wilson's Chlorination Process..12mo, 1 50
Cyanide Processes..12mo, 1 50

MINERALOGY.

Barringer's Description of Minerals of Commercial Value.......Oblong, mor. 2 50
Boyd's Resources of Southwest Virginia..................................8vo, 3 00
Boyd's Map of Southwest Virginia.........................Pocket-book form. 2 00
* Browning's Introduction to the Rarer Elements.........................8vo, 1 50
Brush's Manual of Determinative Mineralogy. (Penfield)..................8vo, 4 00
Butler's Pocket Hand-Book of Minerals...............................16mo, mor. 3 00
Chester's Catalogue of Minerals..................................8vo, paper, 1 00
Cloth, 1 25
* Crane's Gold and Silver...8vo, 5 00
Dana's First Appendix to Dana's New "System of Mineralogy."..Large 8vo, 1 00
Manual of Mineralogy and Petrography...................................12mo 2 00
Minerals and How to Study Them...12mo, 1 50
System of Mineralogy..........................Large 8vo, half leather, 12 50
Text-book of Mineralogy...8vo, 4 00
Douglas's Untechnical Addresses on Technical Subjects..................12mo, 1 00
Eakle's Mineral Tables..8vo, 1 25
Stone and Clay Products Used in Engineering. (In Preparation.)

Egleston's Catalogue of Minerals and Synonyms. 8vo, 2 50
Goesel's Minerals and Metals: A Reference Book. . . . . .16mo mor. 3 00
Groth's Introduction to Chemical Crystallography (Marshall) 12mo, 1 25
* Iddings's Rock Minerals 8vo, 5 00
Johannsen's Determination of Rock-forming Minerals in Thin Sections. 8vo, 4 00
* Martin's Laboratory Guide to Qualitative Analysis with the Blowpipe. 12mo, 60
Merrill's Non-metallic Minerals: Their Occurrence and Uses 8vo, 4 00
Stones for Building and Decoration. 8vo, 5 00
* Penfield's Notes on Determinative Mineralogy and Record of Mineral Tests.
8vo, paper, 50
Tables of Minerals, Including the Use of Minerals and Statistics of Domestic Production. 8vo, 1 00
* Pirsson's Rocks and Rock Minerals 12mo, 2 50
* Richards's Synopsis of Mineral Characters. . . . . .12mo, mor. 1 25
* Ries's Clays: Their Occurrence, Properties, and Uses. 8vo, 5 00
* Tillman's Text-book of Important Minerals and Rocks. 8vo, 2 00

MINING.

* Beard's Mine Gases and Explosions. . . . . .Large 12mo, 3 00
Boyd's Map of Southwest Virginia. . . . . .Pocket-book form, 2 00
Resources of Southwest Virginia. 8vo, 3 00
* Crane's Gold and Silver 8vo, 5 00
Douglas's Untechnical Addresses on Technical Subjects. 12mo, 1 00
Eissler's Modern High Explosives. 8vo, 4 00
Goesel's Minerals and Metals: A Reference Book . . . . .16mo, mor. 3 00
Ihlseng's Manual of Mining. 8vo, 5 00
* Iles's Lead-smelting. 12mo, 2 50
Miller's Cyanide Process. 12mo, 1 00
O'Driscoll's Notes on the Treatment of Gold Ores. 8vo, 2 00
Peele's Compressed Air Plant for Mines 8vo, 3 00
Riemer's Shaft Sinking Under Difficult Conditions. (Corning and Peele) . . .8vo, 3 00
Robine and Lenglen's Cyanide Industry. (Le Clerc). 8vo, 4 00
* Weaver's Military Explosives. 8vo, 3 00
Wilson's Chlorination Process. 12mo, 1 50
Cyanide Processes. 12mo, 1 50
Hydraulic and Placer Mining. 2d edition, rewritten 12mo, 2 50
Treatise on Practical and Theoretical Mine Ventilation. 12mo, 1 25

SANITARY SCIENCE.

Association of State and National Food and Dairy Departments, Hartford Meeting, 1906 . . . . .8vo, 3 00
Jamestown Meeting, 1907. . . . . .8vo, 3 00
* Bashore's Outlines of Practical Sanitation. 12mo, 1 25
Sanitation of a Country House. 12mo, 1 00
Sanitation of Recreation Camps and Parks. 12mo, 1 00
Folwell's Sewerage. (Designing, Construction, and Maintenance). 8vo, 3 00
Water-supply Engineering. 8vo, 4 00
Fowler's Sewage Works Analyses. 12mo, 2 00
Fuertes's Water-filtration Works. 12mo, 2 50
Water and Public Health. 12mo, 1 50
Gerhard's Guide to Sanitary House-inspection 16mo, 1 00
* Modern Baths and Bath Houses 8vo, 3 00
Sanitation of Public Buildings. 12mo, 1 50
Hazen's Clean Water and How to Get It. . . . . .Large 12mo, 1 50
Filtration of Public Water-supplies. 8vo, 3 00
Kinnicut, Winslow and Pratt's Purification of Sewage. (In Press.)
Leach's Inspection and Analysis of Food with Special Reference to State Control. 8vo, 7 00

Mason's Examination of Water. (Chemical and Bacteriological)......12mo, 1 25
Water-supply. (Considered Principally from a Sanitary Standpoint)..8vo, 4 00
* Merriman's Elements of Sanitary Engineering..................8vo, 2 00
Ogden's Sewer Design...............................12mo, 2 00
Parsons's Disposal of Municipal Refuse.......................8vo, 2 00
Prescott and Winslow's Elements of Water Bacteriology, with Special Reference to Sanitary Water Analysis....................12mo, 1 50
* Price's Handbook on Sanitation.........................12mo, 1 50
Richards's Cost of Cleanness. A Twentieth Century Problem 12mo, 1 00
Cost of Food. A Study in Dietaries........................12mo, 1 00
Cost of Living as Modified by Sanitary Science................12mo, 1 00
Cost of Shelter. A Study in Economics.....................12mo, 1 00
* Richards and Williams's Dietary Computer....................8vo, 1 50
Richards and Woodman's Air, Water, and Food from a Sanitary Standpoint.....................................8vo, 2 00
Rideal's Disinfection and the Preservation of Food...............8vo, 4 00
Sewage and Bacterial Purification of Sewage...............8vo, 4 00
Soper's Air and Ventilation of Subways...................Large 12mo, 2 50
Turneaure and Russell's Public Water-supplies..................8vo, 5 00
Venable's Garbage Crematories in America.....................8vo, 2 00
Method and Devices for Bacterial Treatment of Sewage............8vo, 3 00
Ward and Whipple's Freshwater Biology.......................12mo, 2 50
Whipple's Microscopy of Drinking-water.......................8vo, 3 50
* Typhod Fever...............................Large 12mo, 3 00
Value of Pure Water............................Large 12mo, 1 00
Winslow's Bacterial Classification.........................12mo, 2 50
Winton's Microscopy of Vegetable Foods......................8vo, 7 50

MISCELLANEOUS.

Emmons's Geological Guide-book of the Rocky Mountain Excursion of the International Congress of Geologists....................Large 8vo, 1 50
Ferrel's Popular Treatise on the Winds.......................8vo, 4 00
Fitzgerald's Boston Machinist............................18mo, 1 00
Gannett's Statistical Abstract of the World...................24mo, 75
Haines's American Railway Management.....................12mo, 2 50
* Hanusek's The Microscopy of Technical Products. (Winton).......... 8vo, 5 00
Owen's The Dyeing and Cleaning of Textile Fabrics. (Standage). (In Press.)
Ricketts's History of Rensselaer Polytechnic Institute 1824-1894.
Large 12mo, 3 00
Rotherham's Emphasized New Testament....................Large 8vo, 2 00
Standage's Decoration of Wood, Glass, Metal, etc.................12mo, 2 00
Thome's Structural and Physiological Botany. (Bennett).............16mo, 2 25
Westermaier's Compendium of General Botany. (Schneider)............8vo, 2 00
Winslow's Elements of Applied Microscopy....................12mo, 1 50

HEBREW AND CHALDEE TEXT-BOOKS.

Green's Elementary Hebrew Grammar.........................12mo, 1 25
Gesenius's Hebrew and Chaldee Lexicon to the Old Testament Scriptures. (Tregelles)..............................Small 4to, half mor. 5 00

BV - #0027 - 250823 - C0 - 229/152/15 - PB - 9781332047147 - Gloss Lamination